Transcription Factors

The Practical Approach Series

SERIES EDITORS

D. RICKWOOD
Department of Biology, University of Essex
Wivenhoe Park, Colchester, Essex CO4 3SQ, UK

B. D. HAMES
Department of Biochemistry and Molecular Biology,
University of Leeds, Leeds LS2 9JT, UK

Affinity Chromatography
Anaerobic Microbiology
Animal Cell Culture (2nd Edition)
Animal Virus Pathogenesis
Antibodies I and II
Biochemical Toxicology
Biological Data Analysis
Biological Membranes
Biomechanics—Materials
Biomechanics—Structures and Systems
Biosensors
Carbohydrate Analysis
Cell–Cell Interactions
Cell Growth and Division
Cellular Calcium
Cellular Neurobiology
Centrifugation (2nd Edition)
Clinical Immunology
Computers in Microbiology
Crystallization of Nucleic Acids and Proteins
Cytokines
The Cytoskeleton
Diagnostic Molecular Pathology I and II
Directed Mutagenesis
DNA Cloning I, II, and III
Drosophila
Electron Microscopy in Biology
Electron Microscopy in Molecular Biology
Electrophysiology
Enzyme Assays
Essential Molecular Biology I and II
Eukaryotic Gene Transcription
Experimental Neuroanatomy
Fermentation
Flow Cytometry
Gel Electrophoresis of Nucleic Acids (2nd Edition)
Gel Electrophoresis of Proteins (2nd Edition)
Genome Analysis
Growth Factors
Haemopoiesis
Histocompatibility Testing
HPLC of Macromolecules
HPLC of Small Molecules
Human Cytogenetics I and II (2nd Edition)
Human Genetic Diseases
Immobilised Cells and Enzymes

Transcription Factors

A Practical Approach

Edited by

DAVID S. LATCHMAN

Professor of Molecular Pathology, University College and
Middlesex School of Medicine, London, UK

IRL PRESS
—at—
OXFORD UNIVERSITY PRESS
Oxford New York Tokyo

Oxford University Press, Walton Street, Oxford OX2 6DP

Oxford   New York
Athens   Auckland   Bangkok   Bombay
Calcutta   Cape Town   Dar es Salaam   Delhi
Florence   Hong Kong   Istanbul   Karachi
Kuala Lumpur   Madras   Madrid   Melbourne
Mexico City   Nairobi   Paris   Singapore
Taipei   Tokyo   Toronto

and associated companies in
Berlin   Ibadan

Oxford is a trade mark of Oxford University Press

A Practical Approach 🛡 is a registered trade mark
of the Chancellor, Masters, and Scholars of the University of Oxford
trading as Oxford University Press

Published in the United States
by Oxford University Press Inc., New York

A catalogue record for this book is available from the British Library

Library of Congress Cataloging in Publication Data

Transcription factors : a practical approach / edited by David S. Latchman.
(The Practical approach series)
Includes bibliographical references and index.
1. Transcription factors.   2. Genetic regulation.   3 DNA-protein
interaction.   I. Latchman, David S.   II. Series.
[DNLM: 1. Transcription Factors—physiology.   QH 450.2 T7717]
QP552.T68T73   1993   574;87'3223—dc20   92–48702
ISBN 0 19 963342 8 (Hbk)
ISBN 0 19 963341 X (Pbk)

Printed in Great Britain by Information Press Ltd, Oxford, England

Preface

As well as being the essential first step in the conversion of the genetic information in the DNA into protein, the process of transcription is also the major point at which gene expression is regulated. Thus, whilst some cases of post-transcriptional control do exist, in most cases gene regulation is achieved by activating (or repressing) the transcription of particular genes in specific cell types or in response to a specific signal. Once this has occurred, all the other stages of gene expression (RNA processing, translation etc.) follow and the appropriate protein is produced in a cell-type specific or inducible manner. Both the basal process of transcription itself and its regulation are controlled by specific short DNA sequences in the gene promoters or enhancers. These sequences act by binding specific proteins known as transcription factors, which then influence the rate of transcription of the gene.

The study of these transcription factors is, therefore, a critical aspect of gene regulation. In general, however, the characterization of these factors involves a distinct set of methods for studying the proteins themselves and their interaction with DNA, which may not be available even in a laboratory skilled in the standard molecular biology techniques for studying DNA and RNA. The aim of this book is to provide such a set of methods which will allow the user who has identified specific regulatory regions in the gene of interest to completely characterize the protein(s) which bind to them. Initially, this study will involve identifying the proteins binding to a specific DNA sequence within the regulatory region by the DNA mobility shift assay (Chapter 1) as well as characterizing the DNA–protein interaction in more detail using DNaseI footprinting and methylation interference techniques (Chapter 2). Subsequently, the biochemical characteristics of the protein can be studied, allowing determination of its size and its ability to form complexes as well as to stimulate transcription *in vitro* (Chapter 3). Although many such studies emphasize the ability of the factor to stimulate transcription, it should not be forgotten that DNA-binding transcription factors are members of a larger class of proteins which have the ability to bind to DNA. Many of the methods developed for these proteins are, therefore, applicable to transcription factors also (Appendix 1).

This characterization of the factor paves the way for its eventual purification, which in turn allows the partial protein sequence to be determined, thereby allowing the isolation of cDNA clones for the factor by screening cDNA libraries with appropriate oligonucleotides (Chapter 4). Other methods of isolating cDNA clones for the factor which may be more convenient in some situations also exist, for example direct screening of cDNA expression libraries with the DNA-binding site for the factor or antibodies to it (Chapter 5) or cloning by homology to other known factors (Chapter 6).

Once the cDNA clones have been isolated, all the standard techniques of molecular biology can be applied to studying the gene structure, expression pattern, and DNA sequence of the transcription factor. It will also be important, however, to characterize the regions of the protein which are responsible for its various properties such as DNA binding or transcriptional activation, and this can readily be achieved using the cDNA clones (Chapter 7). Ultimately, therefore, an appropriate combination of the methods described here will allow the experimenter to move forward from the characterization of a DNA sequence involved in the basal or regulated transcription of a specific gene and to obtain a detailed understanding of the protein(s) binding to this sequence and the manner in which it controls transcription.

Finally, I would like to thank all the contributors for the efforts they have made to render their methods accessible to others, as well as the staff of Oxford University Press for their continuous assistance.

London   D.S.L.
May 1992

Contents

3. Biochemical characterization of transcription factors   49

Austin J. Cooney, Sophia Y. Tsai, and Ming-Jer Tsai

Contents

4. Purification and cloning of transcription factors   81

R. H. Nicolas and G. H. Goodwin

5. Cloning transcription factors from a cDNA expression library   105

Ian G. Cowell and Helen C. Hurst

Appendices

A1 Physical characterization of DNA-binding proteins in crude preparations   181

Min Li and Stephen Desiderio

A2 Suppliers of specialist items   197

Index   199

Contributors

ALAN ASHWORTH
Chester Beatty Laboratories, The Institute of Cancer Research, Fulham Road, London, SW3 6JB, UK.

AUSTIN J. COONEY
Department of Cell Biology, Baylor College of Medicine, 1 Baylor Plaza, Houston, TX 77030, USA.

IAN G. COWELL
ICRF, Oncology Group, MRC Cyclotron Building, Hammersmith Hospital, Du Cane Road, London, W12 0HS, UK.

C. L. DENT
Wellcome Foundation Ltd., Langley Court, Beckenham, Kent, BR3 3BS, UK.

STEPHEN DESIDERO
Department of Molecular Biology and Genetics, Howard Hughes Medical Institute, Johns Hopkins University School of Medicine, 725 N. Wolfe Street, PCTB, Baltimore, MD 21205, USA.

G. H. GOODWIN
Chester Beatty Laboratories, The Institute for Cancer Research, Fulham Road, London, SW3 6JB, UK.

HELEN C. HURST
ICRF, Oncology Group, MRC Cyclotron Building, Hammersmith Hospital, Du Cane Road, London, W12 0HS, UK.

N. D. LAKIN
Division of Molecular Pathology, University College and Middlesex School of Medicine, The Windeyer Building, Cleveland Street, London, W1P 6DB, UK.

D. S. LATCHMAN
Division of Molecular Pathology, University College and Middlesex School of Medicine, The Windeyer Building, Cleveland Street, London, W1P 6DB, UK.

MIN LI
Department of Physiology, Howard Hughes Medical Institute, UCSF, San Francisco, CA, USA.

Contributors

R. H. NICOLAS
Chester Beatty Laboratories, The Institute of Cancer Research, Fulham Road, London, SW3 6JB, UK.

MALCOLM PARKER
Molecular Endocrinology Laboratory, Imperial Cancer Research Fund, PO Box 123, Lincoln's Inn Fields, London, WC2A 3PX, UK.

MING-JER TSAI
Department of Cell Biology, Baylor College of Medicine, 1 Baylor Plaza, Houston, TX 77030, USA.

SOPHIA Y. TSAI
Department of Cell Biology, Baylor College of Medicine, 1 Baylor Plaza, Houston, TX 77030, USA.

ROGER WHITE
Molecular Endocrinology Laboratory, Imperial Cancer Research Fund, PO Box 123, Lincoln's Inn Fields, London, WC2A 3PX, UK.

Abbreviations

BSA	bovine serum albumin
bZIP	basic zipper
CAT	chloramphenicol acetyl transferase
CIP	calf intestinal phosphatase
CoA	coenzyme A
DATP	diallyl tartardiamide
DCC	dextran-coated charcoal
DES	diethylstilbestrol
DMEM	Dulbecco's modified Eagle's medium
DMP	dimethylpimelidate
DMS	dimethyl sulphate
DNase	deoxyribonuclease
DSP	dithiobis (succinimyl propionate)
DTT	dithiothreitol
GST	glutathione-S-transferase
HMG	high mobility group
HSV	herpes simplex virus
IPTG	isopropyl-β-D-thiogalactopyranoside
LDAO	lauryl dimethylamide oxide
NGF	nerve growth factor
NP40	Nonidet P-40
OD	optical density
PAGE	polyacrylamide gel electrophoresis
PBS	phosphate-buffered saline
PCR	polymerase chain reaction
PCV	packed cell volume
p.f.u.	plaque forming units
PMSF	phenyl methyl sulphonyl fluoride
PR	progesterone receptor
PRE	progesterone response element
SDS	sodium dodecyl sulphate
SRF	serum response factor
SV40	Simian virus 40
TFA	trifluoroacetic acid
Tris	Tris (hydroxymethyl)–aminomethane

1

The DNA mobility shift assay

C. L. DENT and D. S. LATCHMAN

1. Introduction

This chapter presents the DNA mobility shift assay as a method for the identification and investigation of DNA-binding proteins. A knowledge of the DNA-binding protein content of a cell or tissue type can be an important aid towards a more thorough understanding of the functional role of that tissue. Important processes, such as the development of an adult organism from the single-celled zygote and the maintenance of the developed structure and biochemical characteristics of tissues, are increasingly being demonstrated to be dependent on the presence of certain DNA-binding transcription factors in different tissues. The spectrum of transcription factors present determines what genes may be transcribed in a cell type, including genes that encode further transcription factors. The pattern of gene expression is very complex: the expression of a single gene may be determined by a number of different transcription factors, and the ratio of their concentrations may be vital, particularly if competing positive and negative factors are involved. Factors other than those that bind the DNA may be involved, such as protein co-factors, metal ions, and ligand molecules, and a further important factor is the chromatin structure of the DNA. The gene must be available for binding of factors and subsequent transcription.

The mobility shift assay attempts to determine the potential for a gene to be transcribed in a particular cell type by providing an assay for the presence of DNA-binding proteins capable of binding to a promoter. This presents a very simple situation compared with what may be happening to the gene within the context of the chromatin *in vivo*, but, nevertheless, provides a very effective point of first entry into a more detailed understanding. It is possible to correlate directly differences in the protein content with differences in cell behaviour. A simple example where the presence or absence of a single protein appears to determine a cellular characteristic has been dissected in this laboratory (1). Protein extracts from cell types permissive and non-permissive to lytic infection by herpes simplex virus 1 (HSV 1) were assayed for proteins binding to the HSV 1 immediate–early gene promoter TAATGARAT sequence. (The TAATGARAT sequence is a promoter element that has been demonstrated to be important in immediate-early gene transcription via the binding of a complex of a cellular transcription factor protein, Oct-1, and a viral component, Vmw65—see references 2 and 3). Both permissive and non-permissive cell types contain this cellular factor, however, non-permissive cell types were found also to contain a second factor believed to be a neuronal form of the B-cell-specific protein Oct-2, which is closely related in DNA-binding specificity and structurally to Oct-1. However, Oct-2 cannot interact with the viral component Vmw65 (4) and cannot act as a transcriptional activator from the viral TAATGARAT sequence. It is, however, able to bind the TAATGARAT, thus preventing the access of Oct-1/Vmw65 complexes, and subsequent transcription of the immediate early genes of HSV 1. *Figure 2* shows DNA mobility shift assay demonstrating the presence of DNA-binding proteins in the permissive (BHK) and non-permissive (neuronally derived ND7) cells showing the presence of large amounts of the second protein (lower band) in the non-permissive cell line. Further experiments have supported the fact that this protein is a neuronal form of Oct-2 and does indeed have a repressive capacity (5), thus demonstrating how a single band from a mobility shift may eventually explain an important cellular phenomenon.

A detailed description of the techniques involved in the use of the DNA mobility shift assay will be given in this chapter, including methods for the preparation of both nuclear and whole-cell protein extracts. A comparison of the advantages of using each type of extract will be made. We shall also describe other methods for obtaining protein for the assay, including the preparation of mini-extracts (when tissue is limiting) and methods for expressing cloned transcription factors for use in these assays. Specialized approaches for the determination of DNA sequence binding specificity and further characterization of the binding factor will also be discussed.

2. Detection of DNA-binding proteins

2.1 Applications of the DNA mobility shift assay

Sequence specific DNA-binding proteins have been identified in all cell types examined. These are involved in the regulation of transcription and DNA replication. The DNA mobility shift assay provides a powerful tool for the detection of factors binding to specific sequences. The method relies on the ability of a protein to bind to a radiolabelled DNA fragment *in vitro*, followed by electro-phoretic separation of DNA–protein complexes from the un-bound DNA on non-denaturing polyacrylamide gels (6, 7). One or more proteins binding to the DNA fragment may be identified. In general, the larger the protein, the greater the extent of retardation of mobility within the gel. The theory behind the DNA mobility shift assay is illustrated in *Figure 1*. *Figure 2* illustrates a typical example of such an assay. The unbound probe containing the octamer is represented by the very heavy broad band at the bottom of the photograph in all four tracks. Two of the tracks show the

2

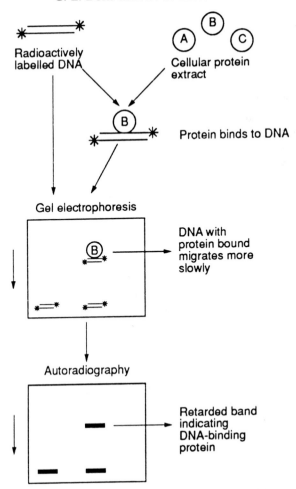

Figure 1. DNA mobility shift assay. Binding of a cellular protein (B) to the radioactively labelled DNA causes it to move more slowly upon gel electrophoresis and hence results in the appearance of a retarded band upon autoradiography to detect the radioactive label.

complexes formed by two DNA-binding proteins of different mobilities binding to the same DNA sequence in extracts prepared from different cell types. The second track shows only one protein, whereas the fourth track shows two. The protein present in both tracks is Oct-1 and has a molecular weight of 100 kDa. The protein present only in the fourth track is Oct-2 and has previously been demonstrated to be smaller than Oct-1 (molecular weight 60 kDa). The complex formed by the smaller protein migrates further within the gel, as would be expected. An accurate determination of the molecular weight of a DNA-binding protein is not possible on this type of gel. The protein must be somehow purified (for example by cutting out of a gel after

3

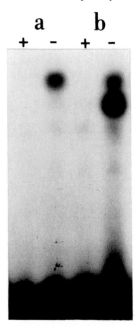

Figure 2. DNA mobility shift assay using nuclear extracts prepared from BHK cells (a) or ND cells (b) and a labelled octamer oligonucleotide. The tracks show the results in the absence (−) or presence (+) of a 100-fold excess of unlabelled octamer oligonucleotide competitor. For further details see reference 1.

ultraviolet cross-linking to the labelled DNA probe) and then run on an SDS-denaturing polyacrylamide gel (see Chapter 3). The DNA sequence specificity of the complex can be tested by competing for binding with non-radiolabelled DNA fragments. If a molar excess of a DNA fragment capable of binding the same protein is introduced into the binding reaction, much of the protein will bind to the unlabelled DNA, leaving less protein available for binding to the probe. This will lead to a reduction in, or elimination of, the band corresponding to the complex formed by that protein (*Figure 3*). In *Figure 2*, the first track represents the same binding reaction as the second track, but in the presence of a 100-fold excess of unlabelled oligonucleotide. The band formed by the interaction between the probe and Oct-1 is competed away. Similarly, the third track represents the same reaction as the fourth track in the presence of excess unlabelled oligonucleotide. The band formed by the interaction between the probe and Oct-1 is competed away. Both specific and non-specific DNA sequences should be used because the failure of a non-specific sequence to compete provides proof that the complex is indeed DNA sequence specific. Identification of proteins within the complex may be made by including antibodies against known proteins in the binding reaction. These antibodies may bind to the complex, causing further electrophoretic retardation and

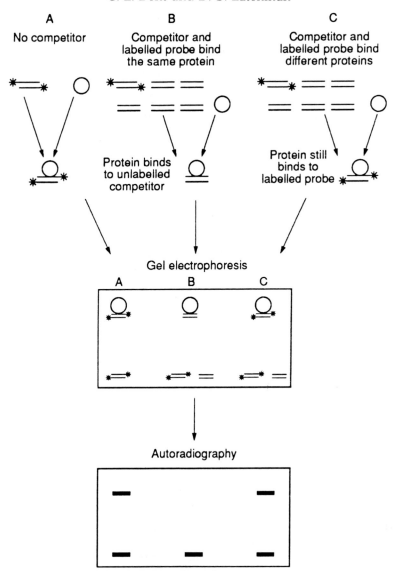

Figure 3. Use of unlabelled competitor DNAs in the DNA mobility shift assay. If the unlabelled competitor is capable of binding the same protein as the labelled probe, it will do so (B) and the retarded band will not be observed.

'super-shifting' the complex, or completely inhibit complex formation by binding to a vital site within the binding protein. These types of experiment will be fully discussed later in this chapter (Section 5.1).

Applications of the mobility shift assay include the identification of both known and novel factors binding to a candidate DNA fragment, usually the

DNA sequences 5′ to a transcription unit. It may also be used to identify fluctuations in the levels of known transcription factors in response to stimuli, e.g. growth factors. In our laboratory, we have used the assay both for the study of known factors and identification of new factors. In one series of experiments (8), we have taken a sequence, identified within the human papillomavirus 16 enhancer as related to the octamer, and used this sequence to identify a protein contained in cervical cell extracts that binds to this sequence. *Figure 4* shows a band-shift carried out using the papillomavirus octamer sequence, showing binding to Oct-1 in all seven cell lines studied, and to a second protein with a higher mobility only in those extracts made from cervical cell lines. In another series of experiments, we have studied variations in the levels of the previously characterized protein Oct-2 on treatment of dorsal root ganglion cells (in culture) with nerve growth factor (NGF). In response to NGF, the level of Oct-2 was shown to increase two- to threefold.

It is important to realize, however, that this assay does have limitations,

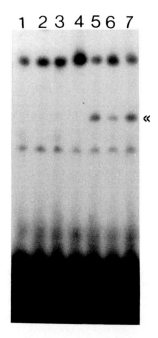

Figure 4. DNA mobility shift assay using an overlapping octamer/TAATGARAT oligo-nucleotide (ATGCTAATGAGAT) from the HSV-1 IE1 gene promoter and extracts from 3T3 cells (track 1), BHK-21 cells (track 2) Jurkat T cells (track 3), S115 mammary epithelial cells (track 4), 310 primary cervical cells with no evidence of papillomavirus infection (track 5), 310 A cells (310 cells transformed with HPV-16 DNA, track 6), and SiHa cervical carcinoma cells (track 7). The arrow indicates the cervical-specific band. For further details see reference 8.

which shall be discussed. As with many techniques, it is at its most valuable when used in conjunction with the other binding and functional assays discussed in the accompanying chapters of this book.

2.2 Selection of DNA probe

A restriction fragment or synthetic oligonucleotide probe may be used, but the size of the fragment is normally kept below about 250 base pairs to enable clear distinction of the probe from any complexes. The type and size of probe used depends on the nature of the investigation. If a previously identified factor is to be studied, then an oligonucleotide probe should be used. An oligonucleotide probe simplifies the interpretation of results as a site is isolated from other unidentified possible sites present within the same regulatory region. However, interactions of a protein on an isolated site may not mirror the situation *in vivo* as competition and co-operation between proteins binding to adjacent sites may be observed. An example is in the U2 small nuclear RNA enhancer where Oct-1 and Sp-1 bind co-operatively to adjacent binding sites (9). In this case, oligonucleotides covering both sites were constructed, and their binding of the two proteins compared with oligonucleotides in which one or the other site was rendered non-functional by mutation.

The gel retardation assay is often used to map the sites in the 5′ enhancer/promoter region of a cloned gene. In this case, a larger restriction fragment probe should be selected, covering the region where regulatory proteins may be expected to bind. Sequence analysis will enable the identification of potential binding sites for known factors; the gel retardation assay may then be used to determine if these factors bind and to identify the binding of novel proteins to unidentified sites. The fragment may then be subjected to deletion analysis, enabling a more precise determination of the position of binding sites. Competition analysis with unlabelled oligonucleotides for defined sites can then be used to identify known sites within the fragment involved in binding.

Protocols 1–3 below give methods for labelling both oligonucleotide and fragment probes.

2.3 Preparation of labelled oligonucleotide probes for retardation assay

Synthetic binding sites are made as two complementary single-stranded oligonucleotides, and these must first be made double stranded by annealing both strands. Annealing is achieved by mixing equimolar amounts of the two oligonucleotides, heating to 80 °C for 5 min, and allowing the oligonucleotides to cool slowly down to room temperature. The oligonucleotides may be designed to possess the overhanging ends of a restriction enzyme site when annealed. This permits them to be cloned into promoter constructs facilitating the assay of their activity in cells.

The easiest method for the labelling of oligonucleotides is to add a ^{32}P-labelled phosphate to the 5′ end using T4 DNA kinase (see *Protocol 1*). Fragment probes may also be labelled using this method but the terminal phosphate must be removed from the DNA by phosphatase treatment before the addition of labelled phosphate (*Protocol 2*). It is usually simpler to label fragment probes by filling in the recessed ends created by restriction enzymes. The methods for labelling restriction enzyme fragments are described in Section 2.4. All the methods for labelling DNA are described in Sambrook *et al.* (10).

Protocol 1. End-labelling DNA with T4 kinase

1. Mix 2 pmol of annealed oligonucleotide with 20 μCi [^{32}P]ATP in the presence of 50 mM Tris–HCl (pH 7.6), 10 mM MgCl$_2$, 5 mM dithiothreitol (DTT), 0.1 mM EDTA, and 0.5 μl (5 units) of T4 DNA kinase.

2. Incubate at 37°C for 30 min.

3. Add 200 μl of STE (10 mM Tris–HCl (pH 8.0), 100 mM NaCl, 1 mM EDTA) to the reaction and separate the labelled oligonucleotide from unincorporated label by centrifugation through a 1 ml Sephadex G-25 column. The probe passes through in approximately 200 μl of which 1 μl (10 fmol of DNA) is used per binding reaction.

Protocol 2. Dephosphorylation of DNA

1. Dissolve DNA in minimum volume of 10 mM Tris–HCl (pH 8.0).

2. Add 5 μl 10 × CIP buffer (0.5 M Tris–HCl (pH 9.0), 10 mM MgCl$_2$, 1 mM ZnCl$_2$, 10 mM spermidine) and 0.01 units of calf intestinal phosphatase per picomole of DNA ends, and make up to 50 μl with distilled water.

3. Incubate at 37°C for 30 min.

4. Add 50 μl H$_2$O, 10 μl STE (100 mM Tris–HCl (pH 8.0), 1 M NaCl, 10 mM EDTA) and 5 μl 10% SDS.

5. Heat to 68°C for 15 min.

6. Extract the DNA with an equal volume of phenol/chloroform (phenol:chloroform:isoamylalcohol, 25:24:1).

7. Take the upper aqueous layer and repeat step 6.

8. Transfer the upper aqueous layer to a fresh tube and extract with an equal volume of chloroform.

9. Take the aqueous layer and precipitate the DNA with 2 volumes of ethanol overnight at −20°C.

2.4 Labelling of fragment probes

There are three classes of fragment probes, depending on the nature of DNA ends generated by the enzymes used for their isolation. The enzymology of labelling these fragments varies, depending on the type of ends generated.

(a) Fragments with 5′ overhanging ends, generated by enzymes such as *Bam*H1 and *Eco*R1. These fragments are the most straightforward to label using the Klenow fragment of DNA polymerase I; therefore, usually enzymes that generate these ends are chosen preferentially when deciding upon a strategy for the isolation of DNA fragments for labelling.

(b) Fragments with 3′ overhanging ends, generated by enzymes such as *Kpn*1 and *Sst*1. These are labelled using T4 DNA polymerase which possesses a 3′ to 5′ polymerase activity as well as a potent 5′ to 3′ exonuclease activity (which must be kept inactive to prevent unwanted degradation of the DNA fragment).

(c) Blunt-ended fragments, generated by enzymes such as *Sma*1 and *Rsa*1. These are also labelled by T4 DNA polymerase after limited digestion by the exonuclease activity of the enzyme.

Although both enzymes have been used in the generation of probes, the Klenow fragment is a more reliable enzyme than T4 DNA polymerase, and should therefore be used if possible. Methods for labelling using both enzymes are given in *Protocols 3* and *4*.

Protocol 3. Filling in 5′ overhangs using the Klenow fragment of *E. coli* DNA polymerase I

1. Mix DNA, 20 μCi [^{32}P]dCTP, 1 mM unlabelled dATP, TTP, and dGTP, 50 mM Tris–HCl (pH 7.5), 10 mM $MgSO_4$, 0.1 mM DTT, and 1 μl (1 unit) Klenow enzyme.
2. Incubate at 37°C for 30 min.
3. Make volume up to 200 μl with STE (see *Protocol 1*) and centrifuge through a Sephadex G-50 column (as described in *Protocol 1*).

Protocol 4. Filling in 3′ overhangs using T4 DNA polymerase

1. Mix DNA, 20 μCi [^{32}P]dCTP, 1 mM unlabelled dATP, TTP, dGTP, 33 mM Tris acetate (pH 7.9), 66 mM K acetate, 10 mM Mg acetate, 0.5 mM DTT, 0.1 mg/ml bovine serum albumin (BSA) and 1 μl (2.5 units) T4 DNA polymerase.
2. Incubate at 37°C for 30 min.

Protocol 4. *Continued*

3. Make volume up to 200 µl with STE (see *Protocol 1*) and centrifuge through a 1 ml Sephadex G-50 column (as described in *Protocol 1*).

Blunt-ended fragments are labelled by a modification of *Protocol 4*. The DNA is incubated with T4 DNA polymerase in the absence of dNTPs for about 1 min, allowing the exonuclease activity of the polymerase enzyme to cut back. If a single dNTP is included during this incubation, the exonuclease will only cut back until it reaches that base in the fragment, thus controlling the digestion. The label and remaining dNTPs are then added, and the digested DNA is resynthesized, incorporating the labelled dNTP.

2.5 Preparation of protein extracts

Protein extracts may be prepared from whole cells or isolated nuclei. There are advantages to using both types of extract, or even a combination, and comparing results obtained with both types. The preparation of nuclear extracts results in the isolation of only those binding factors with access to the DNA. Factors isolated from the nucleus will thus have the potential to bind to sites on the chromosomal DNA. However, it must be considered that much of the chromatin is masked by histones and other DNA-binding proteins, thus rendering it inaccessible to transcription factors, hence a binding site *in vitro* may not necessarily represent a binding site *in vivo*.

The preparation of extracts from whole cells enables the entire DNA-binding protein content of the cell to be examined. Some proteins may be present in the cytoplasm rather than the nucleus, and can be identified by the comparison of the binding profiles of nuclear and cytoplasmic extracts. Proteins present in the cytoplasm cannot be involved in transcriptional regulation at that time, due to their lacking access to the chromatin, but are likely to be on 'standby', ready for a quick transport into the nucleus following a certain stimulus. This transport of transcription factors from the cytoplasm to the nucleus may be studied by comparing whole-cell and nuclear extracts before and after treatment of the cells. It is possible that the protein might be involved in interactions with other proteins in the nucleus or cytoplasm, which might help to keep the factor in the right compartment of the cell or keep it in an active or inactive form.

Whole-cell extracts are easier to prepare, requiring less steps in their preparation, which makes their use favourable when a tissue sample is limiting because fewer manipulations present fewer stages at which protein will be lost or damaged. It is also necessary to prepare whole-cell extracts when a sample has been frozen. Freezing damages the nuclear membrane, preventing the preparation of intact nuclei, and thus the preparation of nuclear extracts.

The method for the preparation of nuclear extracts is given in *Protocol 5*. This method is a modification of that described by Dignam *et al.* (11). Whole-

10

cell extracts are made by a modification of this protocol (12) which shall be discussed later, along with modifications required for the preparation of extracts from certain types of tissue, as opposed to cell lines.

Protocol 5. Preparation of nuclear protein extracts

Samples should be kept on ice at all times, centrifugation should be carried out at 4°C.

1. Harvest cells (5×10^7 to 1×10^8) and centrifuge at $250 \times g$ for 10 min.

2. Wash with phosphate-buffered saline (PBS; 100 mM NaCl, 4.5 mM KCl, 7 mM Na_2HPO_4, 3 mM KH_2PO_4). Centrifuge at $250 \times g$ for 10 min.

3. Resuspend in 5 volumes buffer A (10 mM Hepes (pH 7.9), 1.5 mM $MgCl_2$, 10 mM KCl, 0.5 mM DTT, 0.5 mM phenylmethylsulphonyl fluoride (PMSF)).

4. Incubate on ice for 10 min. Centrifuge at $250 \times g$ for 10 min.

5. Resuspend in 3 volumes buffer A. Add Nonidet P-40 (NP40) to 0.05% and homogenize with 20 strokes of a tight-fitting Dounce homogenizer to release the nuclei.

6. Successful release of nuclei may be checked by phase-contrast microscopy (keep a sample before addition of NP40 and homogenization for comparison).

7. Centrifuge at $250 \times g$ for 10 min to pellet the nuclei.

8. Resuspend the pellet in 1 ml buffer C (5 mM Hepes (pH 7.9), 26% glycerol (v/v), 1.5 mM $MgCl_2$, 0.2 mM EDTA, 0.5 mM DTT, 0.5 mM PMSF). Measure total volume and add NaCl to a final concentration of 300 mM. Mix well by inversion.

9. Incubate on ice for 30 min.

10. Centrifuge at $24\,000 \times g$ for 20 min at 4°C.

11. Aliquot supernatant and snap-freeze in dry ice/ethanol. Store extracts in aliquots at −70°C.

If fewer than 5×10^7 cells are to be used, the method in *Protocol 5* may be scaled down accordingly. For very small numbers of cells, the mini-preparation method given in *Protocol 8* should be used.

Whole-cell extracts are made by a modification of the method given in *Protocol 5*. Harvest the cells and wash with PBS. Resuspend in 1 ml of buffer C and homogenize with 20 strokes of a tight-fitting homogenizer. Then add NaCl to a final concentration of 300 mM and continue from step 9 of the nuclear extract preparation method.

Tissue samples can be more difficult to homogenize than cultured cells.

Small pieces of soft tissue may be treated as cultured cells, and we have successfully used this method for the isolation of binding proteins from rat dorsal root ganglion and brain tissue. Larger pieces of tougher tissue or frozen samples may have to be treated more drastically to release the protein. It should prove sufficient for most tissues to use a tissue macerator instead of a Dounce homogenizer, however, particularly tough tissues or frozen samples may have to be frozen in liquid nitrogen and ground to a fine powder before homogenization. Samples that have at any stage been frozen may only be used for the preparation of whole-cell extracts. Fresh tissue samples should be dealt with as quickly as possible to prevent degradation by proteases.

2.6 The binding reaction

Conditions for the binding of protein to DNA vary between research groups. We use a 45 min incubation on ice, whereas others favour incubations at room temperature. Either conditions give good results in our hands, but we consistently use incubation on ice in order to maintain an extent of uniformity between different studies. The binding conditions routinely used in our laboratory are given in *Protocol 6*.

Protocol 6. Conditions for binding

1. Make a 20 μl binding reaction by mixing 20 mM Hepes (pH 7.9), 1 mM MgCl$_2$, 4% Ficoll, 0.5 mM DTT, KCl to a final salt concentration of 50 mM (remembering that any extract added is in 300 mM NaCl), 2 μg poly dIdC (Pharmacia), and 1 μl (10 fmol) ^{32}P-labelled probe and protein extract.

2. Incubate on ice for 40 min.

3. Run samples[a] on a 4% polyacrylamide gel (0.25 × TBE). The gel should be pre-run at 150 V for about 2 h before electrophoresis (current will drop from 20–30 mA to approximately 10 mA during this time). The samples are then run for 2.5 h (or until the bromophenol blue marker dye has run about two-thirds of the way down the gel).

4. Dry gel on to filter paper (1 h, 80 °C, with vacuum) and autoradiograph overnight.

[a] Do not add any loading dye to samples, the Ficoll in the binding buffer provides the density required for loading. Bromophenol blue in glycerol may be added to a spare track as marker. Poly dIdC is added to the binding reaction as a non-specific competitor for the binding of any general DNA binding proteins, leaving the probe free to bind sequence-specific binding proteins. The amount of extract required will vary depending on the protein concentration of the extract and the abundance and affinity of the factor to be studied. It is advisable to determine the amount giving the best results experimentally. The concentration of protein in extracts should be determined by the method of Bradford (reference 13, *Protocol 7*), allowing the comparison of equal amounts of protein from different extracts.

Protocol 7. Bradford assay for protein concentration

1. Following transfection, wash the cells with PBS, harvest them and transfer to a 1.5 ml microcentrifuge tube.

2. Add 100 μl of 0.25 M Tris–HCl (pH 7.5) to the cell pellet.

3. Disrupt the cells by freezing and thawing. To freeze–thaw, immerse the tubes in liquid nitrogen for 2 min and then transfer them to a 37°C water bath. Repeat the cycle three times.

4. Pellet the cell debris and save the supernatant to test for protein. Samples may be saved at this point by storage at −20°C.

5. Add 1 ml of dye reagent[a] to 10 μl of each sample.

6. Measure the absorbance of the sample at 595 nm after 15 min.

7. If the absolute concentration of protein is needed, then a standard curve can be constructed using BSA as standard (draw A_{595} vs. [BSA] mg/ml).

[a] Dye reagent: 100 mg Coomassie brilliant blue G, 30 mg SDS, 50 mg 95% (v/v) ethanol, 100 ml 85% (v/v) phosphoric acid. Dilute the mixture to a final volume of 1 litre, using distilled water.

2.7 Preparation of mini-extracts

Several methods for the preparation of mini-extracts when cells or tissue are limiting have been published. Mini-extracts are particularly quick and easy to prepare, and all steps can usually be carried out in a microcentrifuge tube without the need for steps such as homogenization. This eliminates the loss of sample that may occur when it is continually being transferred between tubes. Because of this, mini-extracts are often prepared when many samples are to be compared, for example when comparing many treatments of cells or making a time-course study of the induction of a transcription factor, and the preparation of large-scale extracts would be cumbersome and unnecessary. The method given in *Protocol 8* is for the preparation of mini-whole-cell extracts (14).

Protocol 8. The preparation of mini-whole-cell extracts

1. Harvest the cells into 1 ml PBS in a microcentrifuge tube, centrifuge for 1 min to pellet.

2. Wash with 1 m PBS, centrifuge for 1 min, remove all traces of PBS.

3. Resuspend cell pellet in 100 μl (or less) extraction buffer (20 mM Hepes (pH 7.8), 450 mM NaCl, 0.4 mM EDTA, 0.5 mM DTT, 25% glycerol, 0.5 mM PMSF).

Protocol 8. *Continued*

4. Freeze (dry ice/ethanol bath) and thaw (37°C water bath) three times.

5. Spin for 10 min in a microcentrifuge at 4°C.

6. Use 10 μl of supernatant for each binding reaction.

---

3. Other sources of protein for use in the binding assay

Purified transcription factor expressed from the cloned gene may be used as a substrate for a DNA-binding assay. The expression of a cloned transcription factor can allow the study of the binding of that factor in the absence of other proteins with which it may interact *in vivo*, and, when expressed in bacterial systems, in the absence of some post-translational modifications that may occur, for example phosphorylation. Expression can serve to confirm the identity of a clone by comparing the band formed by the expressed cloned protein with that observed in cellular extracts. It must be remembered, however, that a cloned protein may not necessarily bind in the same manner as the endogenous protein does in a cellular extract, or even bind at all. Several factors may contribute to differences in binding. Other proteins may be required for, or contribute towards, the formation of a complex, resulting in a different binding pattern, or inability to bind. Proteins synthesized in bacterial systems will not have undergone certain post-translational modification that occur in a mammalian cell, the most common of these being phosphorylation (see for example reference 15). One of the greatest advantages of using expressed cloned transcription factor for DNA-binding and expression studies is that it is possible to manipulate specific regions of the transcription factor protein by mutating the cloned gene. Truncations, deletions, and point mutations of the protein can be made. Binding assays and expression assays can then be used to determine which mutations are defective in DNA binding or transcriptional activation (see Chapter 7).

There are several methods of expressing cloned proteins, these include:

(a) expression in bacteria

(b) expression in mammalian cells

(c) *in vitro* synthesis in reticulocyte lysate from an *in vitro* translated mRNA

(d) baculovirus expression systems

3.1 Expression of DNA-binding proteins in bacteria

Expression in bacterial systems is one of the less commonly used systems for the synthesis of transcription factor protein. All the other systems described use eukaryotic cells or cell extracts for the synthesis of protein, and it makes good common sense to synthesize a eukaryotic protein in a eukaryotic system.

This system has been used in a number of cases (see Chapter 3, Section 5.7). There are several factors that must be considered when using bacterial expression vectors for the expression of transcription factors. Firstly, as with any expression vector, the coding sequence must be inserted into the vector in the correct orientation and reading frame, otherwise the RNA made will not be able to be translated into protein. It is also important to use a cDNA clone rather than one which is interrupted by introns, as bacteria are incapable of editing these from the RNA to give a mature translatable product. Proteins that have been expressed in bacteria will not have undergone certain post-translational modifications that would occur in the eukaryotic cell. This may be a problem when experiments are being carried out to determine, for example, whether the transcription factor encoded by a certain cloned piece of DNA encodes the same factor as that giving a certain band in a retardation assay, because proteins lacking certain modifications may not bind properly, interact correctly with other proteins, or even possess an incorrect apparent molecular weight. However, the production of unmodified protein may be an advantage when studying the modification process itself. If the protein expressed in a bacterial system behaves differently than that expressed in a cellular extract or eukaryotic expression system, it would suggest that that protein required post-translational modification for full functional activity. It is then possible to attempt the phosphorylation of proteins *in vitro*, although this is complicated by the wide variety of protein kinases in a cell, only one of which may be capable of phosphorylating a particular protein. However, it is theoretically possible, if the right protein kinase is selected, to phosphorylate a transcription factor *in vitro* and demonstrate how its activity may alter. The converse experiment is also possible, i.e. a phosphorylated protein from a cell extract or eukaryotic expression system may also be dephosphorylated by treatment with phosphatases, and the effect of this treatment on binding activity studied.

As a final comment, it is important to make sure that all experiments involving the expression of transcription factor protein are properly controlled. Firstly, binding assays carried out using extracts expressing a cloned protein must be compared with assays using extracts made from cells expressing the intact vector only, in order to identify any binding of bacterial rather than expressed eukaryotic proteins. However, even when this precaution is taken it is often impossible to rule out the possibility that the expressed protein is binding with the aid of a bacterial cofactor protein (that may or may not be an analogue of a protein that is required *in vivo*), and results must be interpreted whilst bearing this in mind.

3.2 Expression of proteins in mammalian cells

The expression of proteins in mammalian cells is one of the most commonly used means of producing recombinant transcription factor for gel retardation or expression assay analysis (see Chapter 7, Section 2). It is often possible to express the cloned transcription factor in a cell line in which it is usually active, therefore ensuring that all cofactors and modification proteins required for the expression of that factor will be present. If this is done, it will be necessary to devise some means of distinguishing the cloned transcription factor from the endogenous factor present in the cell. This may be relatively easy if the cloned factor is expressed as a fusion protein, or has been deleted or truncated, and is therefore a different size to the endogenous factor, but it is possible that this may constitute a major problem. If this is going to present a problem, or if the nature of the investigation makes such an experiment of interest, it is also possible to express a transcription factor in a mammalian cell line in which it is not usually expressed. It may be possible to induce gene expression patterns typical of one cell type in another in which such patterns are not normally observed. If extracts are then made from these cell lines expressing the ectopic transcription factor protein and used in a mobility shift assay, the typical size of retarded band usually expected for this factor may be observed, but it is possible that this type of expression experiment may not yield the expected result. This is because, in addition to the expressed protein being studied, the cell line in which this protein is being anomalously expressed is unlikely to express an identical set of proteins to the cell line in which it is usually expressed. One or more of these proteins may be required as cofactors (or for post-translational modification) enabling binding of the transcription factor, and may be absent from the cell line in which the factor is being expressed. Thus, binding may not be observed, despite successful expression of the correct transcription factor.

The transcription factor is cloned into an expression vector as already described in the discussion of expression of proteins in bacterial systems. The vector systems are similar to those found for bacterial systems, in that the coding DNA must be inserted in the correct orientation for expression, in the correct reading frame, and that the resulting expression product may be synthesized as a fusion protein. In addition, the vector must contain sequences to specify capping and polyadenylation of the transcribed RNA sequence. Protein expression is under the control of a mammalian promoter (often a mammalian viral promoter) which may be inducible. Such vectors also contain sequences necessary for growth in bacteria, such as the bacterial origin of replication and an antibiotic resistance gene. This is necessary because selection and growth in bacteria are essential during the cloning of the transcription factor coding sequence, and in order for large amounts of plasmid to be grown in bacteria for transfection into mammalian cells. The methods for introducing DNA into mammalian cells by transfection are given in Chapter 7. Extracts can be made from the cell line expressing the transcription factor in order to detect the DNA binding of the transfected factor by use of the DNA mobility shift assay. Mini-preparation methods should initially be used to isolate protein, but it is possible that transiently transfected cells will contain insufficient protein for detection by a mobility shift assay. If this is the case, it is possible to prepare a protein extract from more plates of transfected cells, or to prepare stably transfected cell lines such that a higher proportion of the cells present are expressing the transcription factor protein.

3.3 Expression by *in vitro* transcription and translation

In vitro transcription and translation is another method that is used very widely for the production of recombinant or cloned transcription factor protein (see Chapter 7, Section 3.1 and reference 16). This and transfection into mammalian cells are the two most commonly used approaches, and both have their advantages. The major advantage of this method is that it gives a clean protein product which is likely to be free of contaminating transcription factors. This facilitates the expression of a cloned protein that would be the same size as, and therefore be confused with, an endogenous factor if expressed in a mammalian cell line. cDNA encoding the transcription factor is cloned into a vector which possess the promoter sequence for a viral RNA polymerase, usually SP6 or T7 polymerases are used (17). The bluescript vector is particularly suitable for this purpose as it contains both SP6 and T7 RNA polymerase promoters either side of a multiple cloning site polylinker. This enables the *in vitro* transcription of either sense or antisense RNA from the cloned sequence, depending on which enzyme is used. After checking an aliquot of the RNA by electrophoresis (there should be sufficient RNA to visualize on an agarose slab gel without the necessity for Northern blotting) it can be translated in a cell-free translation system. A good system, particularly for the translation of transcription factor proteins, is the rabbit reticulocyte lysate. Reticulocytes do not contain a nucleus, and, therefore, have no requirement for transcription factors. This makes the interpretation of mobility shift assay results from the cloned transcription factors much simpler in the absence of additional endogenous factor or factors. However, it is still necessary to include an aliquot of translation extract (after incubation without RNA or with antisense RNA in a mobility shift) to make absolutely certain that it does not contain endogenous proteins that are capable of binding to the probe and thus confusing the interpretation of results. It is also possible to label radioactively any translation product with [^{35}S]methionine during translation, allowing the protein to be sized accurately on an SDS-polyacrylamide gel. A disadvantage of using this method arises when other factors are required during complex formation. The translation extract is unlikely to contain cofactors such as those which may be found in the cell types to which the transcription factor is endogenous. This can be overcome by mixing the translation product with a cellular extract, if the translation product has been labelled with [^{35}S]methionine it can be easily distinguished from any endogenous factor. Methods for *in vitro* transcription and translation are given in Chapter 7.

3.4 Expression of cloned transcription factor in baculovirus

Baculovirus vectors are very efficient producers of recombinant protein (see Chapter 7, Section 3.2), the problem in their use is that standard cloning procedures cannot be used with these vectors, with the coding sequence having to be introduced by homologous recombination. Baculovirus is a virus that infects insect cells, and the cloned protein is, therefore, produced in an insect cell line. The requirement for non-standard techniques is off-putting to many researchers but, if this is overcome, the high levels of expression should provide ample reward. Vmw65 has been expressed in a baculovirus system (18), with the result that enough protein was expressed to be clearly visible when infected (with recombinant baculovirus) and mock-infected cell extracts were compared by electrophoresis on an SDS-polyacrylamide gel. In this case, the recombinant protein represented about 3% of the soluble protein extracted from the infected cells, equivalent to 20 mg of recombinant protein from 10^9 infected cells. This protein was used directly for DNA mobility shift studies.

3.5 Purification of transcription factor protein

The purification of transcription factor proteins can be applied both to endogenous transcription factors and to an artificially expressed factor, and can be used to purify (or partially purify) a factor from a cell extract (see Chapter 4). The purified factor may then be used for DNA-binding assays to further study the relationships involved in binding to the probe, or can be used as a substrate for protein sequence analysis as part of a strategy for the isolation of a clone encoding the factor. Purification methods may also be used to prepare an extract depleted in a factor. The depleted extract can be used to dissect the activity of the protein, for example by adding back a cloned transcription factor.

4. Investigation of DNA-binding specificity

Sequence-specific DNA-binding proteins will often be capable of binding to a series of variations on a basic consensus sequence. A binding site that is further removed from the consensus may bind the factor less strongly and, therefore, be a weaker site for transcriptional activation. Differences in the sequence of flanking DNA outside of the core consensus binding site may also affect binding. These differences in affinity of different sites for the same factor will contribute to the differential effects of a single transcription factor on different genes, and can be studied by competition analysis. The approach used for the measurement of the affinity of a transcription factor for different binding sites is discussed below.

If a binding reaction is set up between a radiolabelled oligonucleotide and protein extract, the protein–DNA complexes will form retarded bands on electrophoresis as already explained. If, when the reaction is set up, as well as adding radiolabelled binding site oligonucleotide, a 100-fold molar excess of unlabelled oligonucleotide is added, then the DNA-binding protein will be able to bind the unlabelled oligonucleotide and the radioactive probe. This will result in a decrease in the amount of factor available for binding to the probe by competition from the unlabelled site, and subsequent reduction in the intensity of the retarded band. If, rather than introducing an unlabelled oligonucleotide identical to the probe, an oligonucleotide corresponding to a different site to which the factor cannot bind is added, there should be no difference in the intensity of the retarded band because the competing oligo-nucleotide is not able to bind the factor. This is the standard means by which a transcription factor is demonstrated to bind in a sequence-specific manner (*Figure 3*). This kind of experiment may then be taken a stage further by introducing, as competitor, variations on the same binding site. The new sequences chosen may be related binding sites found in other genes in order to study the comparative affinities of natural binding sites, or they may be mutations of the consensus binding site designed to study the effect of base substitutions on the DNA–protein interaction. If a related site that binds the same factor is introduced at an excess into a binding reaction, the extent of competition should vary from that obtained using the site identical to the probe. A higher affinity binding site would compete more protein away from the labelled site, resulting in a stronger reduction in the intensity of the retarded band. A lower affinity site would be less efficient in competition, leaving more protein for complex formation with labelled oligonucleotide, resulting in less reduction in the retarded band. Some related sequences may be totally unable to bind, resulting in the competed binding reaction giving a retarded band identical to that obtained when no competitor DNA is added (or indeed, competitor DNA for a totally unrelated binding site). The com-parative affinity of the protein for different sites can be determined by adding different amounts of competitor for the sites to a uniform binding reaction, and measuring the amount of competitor required to reduce the level of protein bound to the radioactive probe by a particular amount. For example, if one sequence competed all of the protein off the probe at a 10-fold molar excess of competitor, but another related sequence required a 100-fold ex-cess, it could be concluded that the first sequence comprised the higher affinity binding site for the factor in question.

Experiments of this type have been used in our laboratory to provide evidence that the neuronal and B-cell forms of Oct-2 are not identical (19). The affinities of the proteins for a panel of four oligonucleotide variants of an overlapping octamer/TAATGARAT consensus sequence (*Figure 5*) were measured. A mobility shift assay was carried out on oligonucleotide C and competed with varying amounts (1-, 10-, and 100-fold molar excess) of the

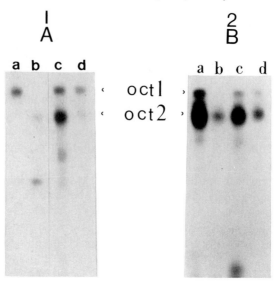

Figure 5. DNA mobility shift assay using four different labelled octamer oligonucleotides (a–d) and extracts prepared from ND7 neuronal cells (A) and Daudi B cells (B). Arrowheads indicate the positions of Oct-1 and Oct-2. Note the different affinities of B cell and neuronal Oct-2 for the different oligonucleotides. For further details see reference 19.

four oligonucleotides and an unrelated Sp1 binding site as a negative control (*Figure 6*). Oct-2 from B cells binds oligonucleotide A with a higher affinity than any other sequence, as demonstrated by the fact that A competes for binding more efficiently. However, Oct-2 from neuronal cells behaves differently, binding to and competing most efficiently with C, thus demonstrating a higher affinity for this sequence.

5. Characterization of DNA-binding proteins

5.1 Addition of antibodies

Antisera raised against a transcription factor may affect its binding by one of two methods. Firstly, the antibody may bind to a site on the transcription factor that is essential for DNA binding, thus totally blocking the ability of the factor to bind DNA and resulting in the complete absence of a DNA–protein complex from the gel. Secondly, if the antibody binds to a non-essential site on the factor, DNA binding may not be impaired, but the mobility of the complex will be altered by the binding of the antibody. The resulting complex will possess a lower mobility than the DNA–protein complex alone, due to the involvement of an additional factor. This will result in a supershift of the DNA–protein complex to a position of lower mobility, due to binding of the antibody.

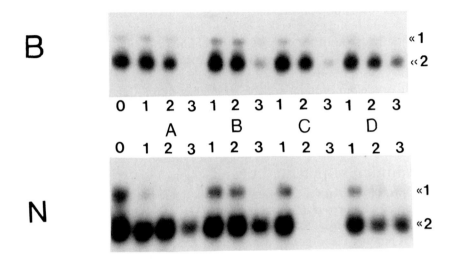

Figure 6. Competition analysis of the B cell (B) and ND cells (N) octamer-binding proteins. DNA mobility shift assays were carried out with labelled oligonucleotide C without competitor (track 0) or in the presence of 1-fold (tracks 1), 10-fold (tracks 2), or 100-fold (tracks 3) excess of unlabelled oligonucleotides A to D (A to D). The arrowheads indicate the positions of Oct-1 (1) and Oct-2 (2). For further details see reference 19.

Antisera may be used to determine whether a mobility shift is caused by a previously identified transcription factor. Evidence such as the size of the shifted band and binding site specificity of a protein could suggest that it may be a previously characterized factor. The binding of an antibody may help towards confirming that the protein is indeed the factor that it is suspected of being (or may demonstrate that it is unique). In one series of experiments, antibodies raised against the transcription factor Oct-1 were found to bind to NFIII, an adenovirus DNA replication factor (20). This provided strong evidence that NFIII and Oct-1 were indeed identical, and was supported by additional experiments demonstrating that the anti-Oct-1 antibodies could block NFIII stimulation of adenovirus DNA replication, and that Oct-1 could functionally substitute for NFIII in this system. This work also serves to demonstrate how the results obtained from the DNA mobility shift assay are particularly valuable when coupled with evidence obtained by other means.

If an antiserum does bind to a factor, it does not necessarily prove that it is identical to the protein to which the antiserum was raised; binding may result from the proteins sharing a common domain that includes the epitope for the antibody. However, antisera may demonstrate that two proteins that appear to be different in a mobility shift assay may be related if they share a domain containing the epitope.

Antibody-binding experiments are carried out as already described in

Protocol 6, with the only modification being that the antisera and protein extract should be mixed together and pre-incubated for about 30 min on ice before the addition of the extract to the binding reaction. When setting up such an experiment, it is also important to include a control binding reaction containing the antisera in the absence of protein extract. This is because serum may contain proteins capable of binding to the DNA probe, resulting in extra shifted bands that may affect the interpretation of results. Similarly, a control should be included containing the extract and an unrelated antibody or pre-immune serum.

5.2 Addition of potential ligands

This approach involves attempting to modulate the binding characteristics of a protein by adding to or removing from the binding reaction certain cofactors that may be required for binding. Such ligands would be added to the standard binding reaction mix described in *Protocol 6*.

Factors that may influence the binding reaction include the presence of ions such as calcium, magnesium, and zinc. These may be removed by the addition of chelating agents such as EDTA and EGTA to the binding reaction mix. The activity of a transcription factor in a depleted extract can yield important information about its structure. Sp1 was demonstrated to be inactive in HeLa extracts that had been depleted of zinc by treatment with EDTA. This suggests that Sp1 probably binds to DNA by a zinc-dependent structure known as the zinc finger. In the same series of experiments, Oct-1 was shown to be fully active in zinc-free extracts, demonstrating that its activity is not dependent on zinc.

Some DNA-binding proteins also act as receptors for endogenous substances, examples being the steroid and retinoic acid receptors. It would thus be obvious in these cases to study the effect of adding the natural ligand to the binding reaction, along with studies involving the addition of ligand analogues which may have either a stimulatory or inhibitory effect compared with the natural ligand (see Chapter 7).

5.3 Proteolytic clipping band-shift assay

The proteolytic clipping band-shift assay can be used alongside antibody binding to examine the relatedness of two proteins. This type of assay involves the addition of various dilutions of a specific protease, e.g. trypsin or chymotrypsin, to a DNA–protein complex, resulting in the production of partial cleavage products that become more completely cleaved with the addition of higher concentrations of protease. The pattern of partial and complete cleavage products gives a typical 'fingerprint' to a protein. If two proteins are identical, they will give identical cleavage products. If two proteins are related, sharing some homologous domains, then some similar cleavage products may be present within a somewhat different fingerprint.

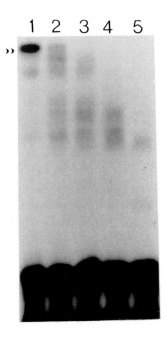

Figure 7. Protease clipping assay in which DNA mobility shifts were carried out with an oligonucleotide which binds the octamer binding protein Oct-1 (arrowed) using an untreated extract (track 1) and the same extract pre-treated with various amounts of chymotrypsin (tracks 2–5).

Figure 7 shows a typical result from a proteolytic clipping band-shift assay, with the first track showing the binding reaction in the absence of protease, and the following tracks showing the cleavage products on addition of increasing amounts of protease.

A method for proteolytic clipping using trypsin and chymotrypsin is given in *Protocol 9*.

Protocol 9. Proteolytic clipping band-shift assay

1. Set up binding reaction exactly as described in *Protocol 6*. Incubate at room temperature for 10 min (not on ice as in *Protocol 6*). Five tubes should be prepared for each type of protease to be included, for four different concentrations of protease and a control without protease.

2. During the binding incubation, prepare serial dilutions of the proteases in water. Keep a stock solution of protease at 4000 units/ml for chymotrypsin and 30 μg/ml for trypsin at 4°C. Working stock solutions of chymotrypsin (0.004, 0.002, 0.001, and 0.0005 units/μl) and trypsin (30, 10, 3, and 1 ng/μl) should be prepared by dilution before use.

Protocol 9. *Continued*

3. Add 1 μl of each dilution of protease to the relevant binding reaction tube, and 1 μl of water to the no-protease control tube. Incubate at room temperature for a further 10 min.

4. Load samples on to a 4% polyacrylamide gel in 0.25 × TBE and electrophorese as described in *Protocol 6*.

---

6. Study of protein–protein interactions

Many DNA binding proteins interact with other proteins during their activity. Factors such as Ap1 are active as a heterodimer of two proteins, Fos and Jun, that will then together recognize and bind to a site in the DNA. A different type of protein cofactor is Vmw65, a HSV 1 protein which binds to Oct-1, forming a complex which can then bind to the HSV-1 immediate–early gene TAATGARAT consensus and enable transactivation of viral RNA synthesis. The Vmw65 protein itself is not capable of binding DNA, although it does contact the DNA sequence when complexed with Vmw65 (2, 3).

The inclusion of additional proteins into DNA-binding complexes may be studied using the gel retardation assay. The binding of an extra protein into a complex will increase its apparent molecular weight on electrophoresis. Vmw65 is by far the most extensively studied cofactor involved in DNA-binding complexes (2, 3). Nuclear extracts made from cells that have been infected with HSV 1 are compared with extracts from mock-infected cells. The mock-infected extract will bind to a TAATGARAT probe, giving a shift typical of the Oct-1 protein, but the infected extract will give a further shift of lower mobility due to the binding of an additional factor, Vmw65, into the complex. *Figure 8* shows the type of mobility shift given by Oct-1 when complexed with Vmw65 to form a larger complex. The shift typical of Oct-1 disappears, and a new, lower mobility shift appears, providing further evidence that Oct-1 is the protein involved in the interaction with Vmw65 (3).

7. Concluding comments

The methods and examples given in this chapter demonstrate that the DNA mobility shift assay is a powerful technique for the identification and study of DNA-binding proteins. It is a basic method, but has many applications and modifications that may yield information about different aspects of transcription factor proteins. However, it must be remembered that the technique does have limitations, and that the demonstration of DNA binding *in vitro* does not necessarily mean that a particular gene is under the control of a particular transcription factor *in vivo*. The following chapters give a detailed

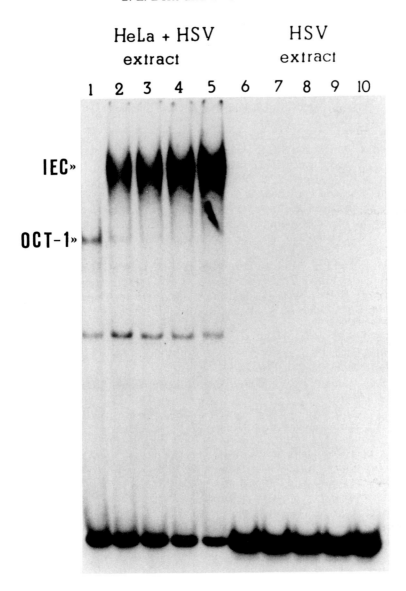

Figure 8. DNA mobility shift assay using a radioactively labelled oligonucleotide containing the octamer-related TAATGARAT sequence. Note that incubation with HeLa cell extract alone results in Oct-1 binding to the probe (track 1). However, addition of increasing amounts of HSV virion extract containing Vmw65 results in the formation of the larger infected extract complex (IEC) containing both Oct-1 and Vmw65 (tracks 2–5). This complex is not formed upon incubation with HSV virion extract alone, indicating that it requires both Oct-1 and Vmw65, and that Vmw65 alone does not bind to DNA (tracks 6–10). Data kindly provided by Dr C. M. Preston. For further details see reference 3.

description of the types of experiment that can, and should, be carried out to support evidence gained using the DNA mobility shift assay.

References

1. Wheatley, S. C., Dent, C. L., Wood, J. N., and Latchman, D. S. (1991). *Exp. Cell. Res.*, **194**, 78.
2. O'Hare, P. and Goding, C. R. (1988). *Cell*, **52**, 435.
3. Preston, C. M., Frame, M. C., and Campbell, M. E. M. (1988). *Cell*, **52**, 425.
4. Gerster, T. and Roeder, R. G. (1988). *Proc. Natl. Acad. Sci. USA*, **85**, 6347.
5. Lillycrop, K. A., Dent, C. L., Wheatley, S. C., Beech, M. N., Ninkina, N. N., Wood, J. N., and Latchman, D. S. (1991). *Neuron*, **7**, 381.
6. Fried, M. and Crothers, D. M. (1981). *Nucleic Acids Res.*, **9**, 6505.
7. Garner, M. M. and Revzin, A. (1981). *Nucleic Acids Res.*, **9**, 3047.
8. Dent, C. L., McIndoe, G. A., and Latchman, D. S. (1991). *Nucleic Acids Res.*, **19**, 4531.
9. Janson, L. and Pettersson, U. (1990). *Proc. Natl. Acad. Sci. USA*, **87**, 4732.
10. Sambrook, J., Fritsch, E. T., and Maniatis, T. (ed.) (1989). *Molecular Cloning, a Laboratory Manual* (second edition), Cold Spring Harbor Press, NY, USA.
11. Dignam, J. D., Lebovitz, R. M., and Roeder, R. G. (1983). *Nucleic Acids Res.*, **11**, 1575.
12. Manley, J. L., Fine, A., Cano, A., Sharp, P. A., and Gefter, M. L. (1980). *Proc. Natl. Acad. Sci. USA*, **77**, 3855.
13. Bradford, M. (1976). *Anal. Biochem.*, **72**, 248.
14. Schreiber, E., Matthias, P., Muller, M., and Schaffner, W. (1989). *Nucleic Acids Res.*, **17**, 6419.
15. Roberts, S. B., Segil, N., and Heintz, N. (1991). *Science*, **253**, 1022.
16. Neuberg, M., Adamkiewicz, J., Hunter, J. B., and Muller, R. (1989). *Nature*, **341**, 243.
17. Green, M. R., Maniatis, T., and Melton, D. A. (1983). *Cell*, **32**, 681.
18. Kristie, T. M., Le Bowitz, J. H., and Sharp, P. A. (1989). *EMBO J.*, **8**, 4229.
19. Dent, C. L., Lillycrop, K. A., Estridge, J. K., Thomas, N. S. B., and Latchman, D. S. (1991). *Mol. Cell. Biol.*, **11**, 3925.
20. Pruijn, G. J. M., van der Vliet, P. C., Dathan, N. A., and Mattaj, I. W. (1989). *Nucleic Acids Res.*, **17**, 1845.
21. Kadonga, J. T., Carner, K. R., Musiarz, F. R., and Tjian, R. (1987). *Cell*, **51**, 1079.

2

Determination of DNA sequences that bind transcription factors by DNA footprinting

N. D. LAKIN

1. Introduction

The development of assay systems that allow the detection of DNA-binding proteins has led to the discovery of a multitude of putative transcription factors. However, preliminary experiments using the band-shift assay technique (Chapter 1) employ relatively large fragments of a gene promoter as a probe for DNA-binding activity. Thus, this technique does not allow the identification of the precise DNA sequence to which a protein binds. In order to supply this information, a series of techniques have been developed that are known as DNA footprinting. DNA footprinting allows the determination of a short protein-binding site within a relatively large DNA fragment, and thus provides an essential step in the characterization of transcription factors.

The general principle of *in vitro* DNA footprinting is based on the cleavage of the DNA molecule with either a chemical reagent or an enzyme. The DNA of interest is radioactively labelled at one end of the molecule and on one strand of the DNA duplex only. Thus, limited degradation of this DNA, such that each molecule of the DNA is cleaved randomly only once or a few times by the chemical or enzyme, results in a ladder of DNA fragments of varying size when the DNA is subjected to denaturing polyacrylamide electrophoresis and detected by autoradiography (see *Figure 1*). Thus, each band in the DNA ladder is representative of a specific nucleotide where the cleavage agent has cut the labelled strand of the DNA. If Maxam and Gilbert sequencing reactions of the same DNA fragment are run alongside the cleavage products, then the specific nucleotide in the DNA sequence which is represented by each band, can be determined.

If, however, the DNA is complexed with a protein(s) before being treated with the cleavage agent, the nucleotides to which the protein is bound will be protected against cleavage. Thus, certain bands in the DNA ladder, representative of where the protein binds the DNA duplex, will not be

DNA sequences that bind transcription factors

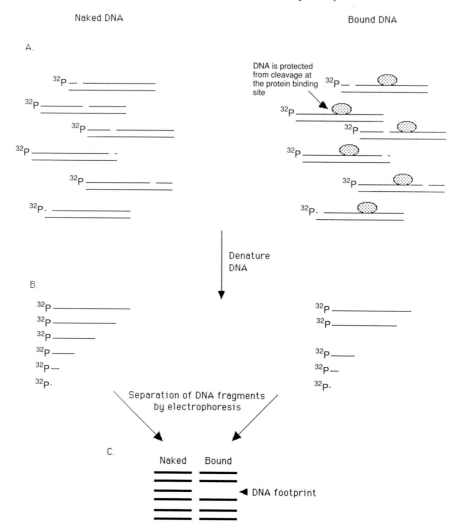

Figure 1. The principle of DNA footprinting. Either naked DNA, or DNA that is bound by protein, is cut randomly on one strand of the DNA duplex by a cleavage agent (A). If the DNA is labelled on one strand of the molecule then, under denaturing conditions, a population of DNA fragments will be produced of a varying size (the size is dependent on the point of cleavage in the DNA molecule) (B). These DNA fragments are then separated by denaturing polyacrylamide gel electrophoresis to produce a ladder of radioactive DNA fragments that can be detected by autoradiography (C, naked DNA). If a protein interacts with the DNA, then the sequence to which it binds is protected against cleavage. This results in the loss of a specific subset of labelled DNA fragments under denaturing conditions (B, bound DNA). Thus, a gap in the DNA ladder (DNA footprint) is observed when this sample is compared with naked DNA (C).

produced due to the protein protecting these nucleotides against cleavage. This results in a gap in the DNA ladder which is representative of a protein binding a specific DNA sequence (see *Figure 1*). By comparing the gap in the ladder to the Maxam and Gilbert sequencing reactions, the precise DNA sequence protected from cleavage, and thus the DNA-binding site of the protein, can be determined.

A number of DNA-footprinting methods that use different agents to cleave the DNA have been developed. This chapter describes the principles of DNA footprinting in its various forms. Included in this are protocols for the most commonly used footprinting methods such as DNase I footprinting, dimethyl sulphate (DMS) protection, and DMS interference footprinting. Furthermore, reference is also given to more recently developed techniques that allow the footprinting of specific DNA fragments *in vivo*.

2. Preparation of DNA fragments labelled on one strand of the DNA duplex

A prerequisite for DNA footprinting is the use of a DNA probe that has been labelled on only one strand of the DNA duplex. Ideally, the region to be footprinted should be 25–100 nucleotides from either end of the DNA fragment used as a probe. This is to ensure that the region of DNA to be investigated is capable of being accurately resolved on a non-denaturing polyacrylamide gel. The labelling of only one strand of a DNA fragment can be achieved in a number of ways, which are described below.

2.1 End-labelling of the DNA fragment using T4 polynucleotide kinase

If convenient restriction sites exist either side of the DNA sequence of interest, then end-labelling of only one strand of the DNA can be achieved by selectively dephosphorylating one strand of the DNA before labelling with polynucleotide kinase (*Protocol 1*). Thus, one end of the DNA fragment is cut with an appropriate restriction enzyme before the 5′ terminal phosphates of the resulting DNA ends are dephosphorylated with calf intestinal alkaline phosphatase (CIP). The DNA is then cut with the second restriction enzyme and the required DNA fragment isolated by agarose gel electrophoresis. When this fragment is labelled with polynucleotide kinase and [γ-^{32}P]ATP, only one strand of the DNA lacks a terminal 5′ phosphate and can, therefore, incorporate the radioactive label.

Protocol 1. Labelling one strand of the DNA duplex using polynucleotide kinase

1. Digest the recombinant plasmid DNA to completion with the appropriate restriction enzyme. Enough plasmid should be digested to yield sufficient

Protocol 1. *Continued*

DNA for end-labelling after isolation of the desired fragment from an agarose gel.

2. Dephosphorylate the 5′ termini of the DNA using CIP (see Chapter 1, *Protocol 12*).

3. Extract the DNA once with phenol:chloroform (1:1) and once with chloroform before precipitating the DNA from the aqueous phase by adding sodium acetate (pH 7.0) to 0.3 M and 2 volumes of ethanol.

4. Resuspend the DNA in 20 μl sterile distilled water and digest to completion with the second restriction enzyme of choice.

5. Separate the resulting DNA fragments by low melting point agarose gel electrophoresis and isolate the required DNA fragment from the gel as described in Sambrook *et al.* (1).

6. Extract the DNA once with phenol:chloroform (1:1) and once with chloroform, add 10 μg of *E. coli* tRNA to the supernatant, and precipitate the nucleic acid as described in step 3.

7. Resuspend the DNA in TE (10 mM Tris–HCl, 1 mM EDTA (pH 8.0)) and determine the concentration by electrophoresing an aliquot of the DNA on an agarose minigel with size standards of a known concentration.

8. Assemble a 20 μl reaction in 1 × kinase buffer (50 mM Tris–HCl (pH 7.6), 10 mM MgCl$_2$, 5 mM DTT, 0.1 mM spermidine) containing 10–20 ng of dephosphorylated DNA fragment, 50 μCi of [γ-^{32}P]ATP and 20 units of T4 polynucleotide kinase. Incubate at 37 °C for 1 h and terminate the reaction by extracting once with phenol:chloroform (1:1) and once with chloroform.

9. Add 10 μg of *E. coli* tRNA and separate the DNA from unincorporated radiolabel by precipitating twice, as described in step 3.

10. Resuspend the labelled DNA to the required concentration in TE (pH 8.0).

2.2 End-labelling using the Klenow fragment of DNA polymerase I

An alternative method for labelling one strand of the DNA probe can be used if only one of the restriction enzymes employed to generate the required DNA fragment creates a 5′ overhang. This can be achieved by filling in the overhang with radiolabelled nucleotides using the large fragment of DNA polymerase I (Klenow fragment). Thus, the DNA is first cut with a restriction enzyme that creates a 5′ overhang and filled in with radioactive nucleotides before being cut with a second restriction enzyme (see *Protocol 2*). This results in both the desired DNA fragment and the vector DNA being labelled at only one end of their respective DNA molecules. The DNA probe can then be separated from the vector DNA by agarose gel electrophoresis, visualized by autoradiography, and isolated from the gel for subsequent footprinting experiments.

Protocol 2. Labelling one strand of the DNA duplex by filling in DNA ends

1. Digest the DNA to completion with the required restriction enzyme that generates a 5′ overhang of the DNA ends.

2. Fill in the DNA ends using the Klenow fragment of DNA polymerase I in conjunction with radioactive nucleotides as described in Chapter 1, *Protocol 3*. Normally, radioactively labelled dCTP and/or dGTP are used in this reaction.

3. Extract the DNA once with phenol:chloroform (1:1) and once with chloroform before adding 10 μg of *E. coli* tRNA to the supernatant and precipitating the nucleic acid with sodium acetate (pH 7.0) and ethanol (see *Protocol 1*, step 3).

4. Resuspend the DNA in sterile distilled water and digest to completion with the second restriction enzyme.

5. DNA can be loaded directly on to a low melting point agarose gel and the resulting DNA fragments separated by electrophoresis.

6. Detect the relevant DNA fragment by autoradiography and isolate from the gel as described by Sambrook *et al.* (1).

7. Extract the DNA once with phenol:chloroform (1:1) and once with chloroform, add 10 μg of *E. coli* tRNA to the supernatant and precipitate the nucleic acid with sodium acetate (pH 7.0) and ethanol (see *Protocol 1*, step 3).

8. Resuspend the DNA to the desired concentration in TE (pH 8.0).

2.3 Amplification of radioactively labelled DNA fragments by the polymerase chain reaction

If no convenient restriction enzyme sites surround the DNA fragment to be footprinted, then DNA fragments can be prepared by exploiting the polymerase chain reaction (PCR). This technique is also advantageous if large quantities of the required DNA sequence are unavailable since only small amounts of template DNA are required.

In brief, the specific DNA fragment required for footprinting is amplified from plasmid DNA by PCR using two gene specific primers. However, one of these primers is radioactively labelled. The result of this is that the amplified DNA fragments are radioactively labelled on one strand of the DNA duplex only. The methodology for this technique has been described elsewhere (2, see Chapter 6).

3. Determination of a transcription factor binding site by DNase I footprinting

Before the development of DNA footprinting techniques, proof of the specificity of a DNA–protein binding reaction ultimately rested on the isolation and characterization of a DNA fragment which was protected from DNase I digestion by the DNA-binding protein (3, 4). However, Galas and Schmitz (5) devised a modification of this technique, named DNase I footprinting, that allowed the specific determination of a protein-binding site within a DNA fragment.

DNase I footprinting uses DNase I as the cleavage agent of the DNA molecules. This technique exploits the property of DNase I that, on limited digestion of a DNA template, it will not completely cleave the DNA duplex but randomly 'nick' one strand of the DNA. Thus, when a population of naked DNA that is labelled on only one strand of the DNA duplex is subjected to a limited digestion with DNase I, it results in a population of DNA molecules that have been cleaved at random only once or a few times on the DNA duplex. Therefore, when the treated DNA is run on a denaturing polyacrylamide gel, a characteristic ladder of DNA fragments is observed. If the DNA is complexed with a protein(s) prior to digestion with DNase I, then certain regions of the DNA will be protected against 'nicking' by this enzyme. This results in the loss of a specific subset of fragments from the DNA population when under denaturing conditions. Thus, when the reaction is compared with control naked DNA reactions, a gap in the DNA ladder produced on electrophoresis is observed. This is the DNA footprint.

Interestingly, some bands adjacent to the protected DNA sequence may appear to be more intense in the presence of the protein. These regions of DNase I hypersensitivity are likely to represent a change in the structure of the DNA when it is bound by proteins, thus rendering it more susceptible to cleavage by DNase I.

To ascertain whether the interaction of a given protein to a fragment of DNA is sequence specific, as opposed to an interaction of a non-specific binding protein with the DNA, a competition reaction is usually performed. This consists of adding a molar excess of unlabelled DNA containing the protein-binding site. This unlabelled DNA is in a molar excess and will therefore sequester the protein and inhibit it from binding the labelled DNA fragment. Thus, the DNase I is now capable of cleaving the labelled DNA at the protein-binding site. This results in a loss of the gap in the DNA ladder that is characteristic of a protein binding to the DNA. When two different DNA-binding sites exist in the footprinted DNA, it is possible to compete away either one or the other footprints by using an oligonucleotide competitor that contains a binding site for only one of the proteins (for example see *Figure 2*).

Generally, the method of DNase I footprinting consists of three steps. In the first of these, the protein (usually contained in a nuclear or whole-cell protein extract) is allowed to complex with the DNA in a similar fashion to that used in the technique of the DNA mobility shift assay (see Chapter 1). Secondly, once this reaction is complete, DNase I is added to the mixture and the reaction allowed to proceed for a short period of time (usually the amount of DNase I added and the period of time the reaction is allowed to proceed has been determined previously using naked DNA in the reaction). Finally, the reaction is terminated and the DNA isolated and analysed by denaturing polyacrylamide gel electrophoresis.

An example of DNase I footprinting is illustrated in *Figure 2*. This example shows two footprints (A and B) produced by two cellular proteins binding distinct sites within the human immunodeficiency virus (HIV) type 1 long terminal repeat negative regulatory element (6). Note that either footprint A or footprint B can be selectively eliminated by the addition of either site A or site B competitor DNA. It is only by the addition of a competitor that contains site A and site B that both the footprints are abolished. *Protocol 3* describes the DNase I footprinting protocol used to produce the data illustrated in *Figure 2*.

Protocol 3. DNase I footprinting

1. Assemble a 100 μl binding reaction in 1 × binding buffer (20 mM Hepes (pH 7.9), 2 mM $MgCl_2$, 50 mM NaCl, 1 mM DTT, 20% glycerol) that contains:
    - 2–5 ng of DNA to be footprinted
    - competitor oligonucleotide (when required)
    - 1–5 μg poly(dI-dC)
    - protein extract

    Incubate on ice for 30 min.[a]

2. Dilute the stock DNase (2 μg/μl) 1:100 in 1 × binding buffer immediately before use and add 1 μl to each sample. Incubate at room temperature for 15–30 sec.[b]

3. Terminate the reaction by adding 100 μl of freshly made stop buffer (50 mM Tris–HCl (pH 8.0), 2% SDS, 10 mM EDTA (pH 8.0), 0.4 mg/ml proteinase K, 100 μg/ml glycogen). Incubate at 37°C for 30 min and then at 70°C for a further 2 min.

4. Extract once with phenol:chloroform (1:1) and once with chloroform before precipitating the DNA from the supernatant under standard conditions of 0.3 M sodium acetate (pH 7.0) and 2 volumes of ethanol.

Protocol 3. *Continued*

5. Resuspend the DNA in 5–10 µl of denaturing loading dye (90% de-ionized formamide, 10 mM EDTA (pH 8.0), 0.01% bromophenol blue, 0.01% xylene cyanol).

6. Heat the samples at 90°C for 3 min. Load 1200–1500 c.p.m. of each sample (as determined by their Cerenkov emission) on a 6% denaturing polyacrylamide:bisacrylamide (19:1) gel containing 1 × TBE (0.09 M Tris–Borate (pH 8.3), 2 mM EDTA) and 42% urea (gels should be pre-run for 30 min prior to loading and then run at 1600 V/30 mA).

7. Fix the gel in 10% acetic acid for 15 min and dry the gel under a vacuum before autoradiography.

[a] The individual requirements for each DNA fragment will have to be established, particularly the amount of poly dI-dC, Mg^{++}, and extract included in the reaction.
[b] The precise time of incubation with DNase to achieve a satisfactory DNA ladder should be determined empirically.

4. Determination of a transcription factor binding site by dimethyl sulphate protection footprinting

The DNase footprinting technique allows the determination of a relatively short DNA sequence that interacts with a transcription factor. Another technique, known as DMS protection footprinting, has been developed to provide further information about the exact bases involved in this interaction. This technique is based on assessing the ability of a transcription factor to protect against the methylation of guanine residues within a DNA sequence by contacting those nucleotides during DNA binding. Thus, in addition to providing the DNA-binding site of a transcription factor, this technique also allows the determination of the precise guanine residues within the DNA sequence that the protein contacts.

Instead of using DMS and piperidine, the DNA may be cleaved using hydroxy radicals. Limited treatment of DNA with hydroxy radicals breaks the DNA backbone with almost no sequence dependence. Therefore, a ladder containing bands that are representative of every nucleotide in the DNA sequence is produced on electrophoresis, such that all the backbone positions of the DNA sequence may be monitored for contact by the protein. This method of footprinting is of particular relevance if the DNA-binding site of a protein contains no guanine residues and has been described elsewhere (7).

The technique of DMS protection footprinting exploits a base-specific chemical cleavage of the DNA that was first developed for sequencing end-labelled DNA fragments (8). In brief, this technique involves the limited

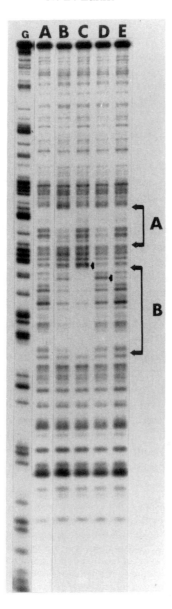

Figure 2. DNase I footprint of protein binding sites A and B of the HIV-LTR negative regulatory element. The HIV-LTR negative regulatory element was labelled on one strand of the DNA duplex and used as the probe in a DNase I footprinting experiment in conjunction with a Jurkat T-cell nuclear extract. A control reaction containing no nuclear extract was also performed (lane A). Lane B contains no competitor, whereas competitor is contained in lane C (site A oligonucleotide), lane D (site B oligonucleotide), and lane E (sites A and B oligonucleotide). Lane G contains a Maxam and Gilbert guanine sequencing reaction of the DNA fragment. Data provided by M. Collins and D. S. Latchman.

treatment of a DNA fragment, which has been labelled on one strand of the DNA duplex, with the methylating agent DMS. This results in the random methylation of one or a few guanine residues in the DNA sequence at position N7 of the purine ring. Subsequent treatment of this partially methylated DNA with piperidine causes the cleavage of the DNA molecules exclusively at guanine residues that have been methylated. This treatment results in the production of a population of different-sized DNA molecules, the length of which is dependent on the specific guanine residue which was methylated and subsequently cleaved by piperidine (see *Figure 3*, naked DNA). Therefore, if the labelled DNA fragments are separated by denaturing gel electrophoresis and detected by autoradiography, a DNA ladder will be produced. The bands within this ladder are representative of the sequence of guanine residues within the DNA fragment.

DMS protection footprinting consists of the end-labelled DNA being complexed with a protein(s) prior to the limited methylation of the DNA by DMS. Thus, DNA is first complexed with proteins by mixing it with a nuclear or whole-cell protein extract under similar conditions to those employed in the DNA mobility shift assay (see Chapter 1). The DNA contained within this reaction is then partially methylated on guanine residues by the addition of DMS to the binding reaction. Following separation by electrophoresis, the band that represents DNA bound by protein and the band produced by unbound DNA are isolated from the gel, and the DNA is cleaved at methylated guanine residues by treatment with piperidine. If a particular guanine residue is contacted by a DNA-binding protein, then it is protected from methylation by DMS. Obviously, therefore, DNA fragments methylated at this residue will be absent from the population of DNA isolated from the bound DNA band. Thus, no cleavage at this nucleotide will occur on treatment with piperidine, resulting in the absence of that specific band from the guanine ladder that is observed on electrophoresis of the DNA (see *Figure 3*,

---

Figure 3. Detection of a protein binding site by DMS protection footprinting. DNA that is labelled on one strand of the DNA duplex is complexed with protein and partially methylated by DMS. Bound and naked DNA fractions are subsequently separated by electrophoresis and the molecules cleaved at methylated guanine residues by piperidine. If the DNA is denatured, a population of labelled DNA fragments of varying length (depending on which of the guanine residues in the sequence was methylated), is produced (B, naked DNA). If these DNA fragments are separated by denaturing polyacrylamide electrophoresis, then a ladder of DNA fragments representative of the sequence of guanine residues will be produced (C, naked DNA). The naked DNA fraction will contain the complete sequence ladder due to all the guanine residues in this sequence being susceptible to methylation by DMS. However, if a protein contacts a specific guanine residue then it will protect it from methylation and subsequent cleavage by piperidine. This results in the bound DNA fraction containing no DNA fragments methylated at this site. Therefore, a specific subset of fragments will be absent from this DNA fraction after cleavage with piperidine (B, bound DNA), resulting in a gap in the guanine ladder when these fragments are separated by electrophoresis (C, bound DNA).

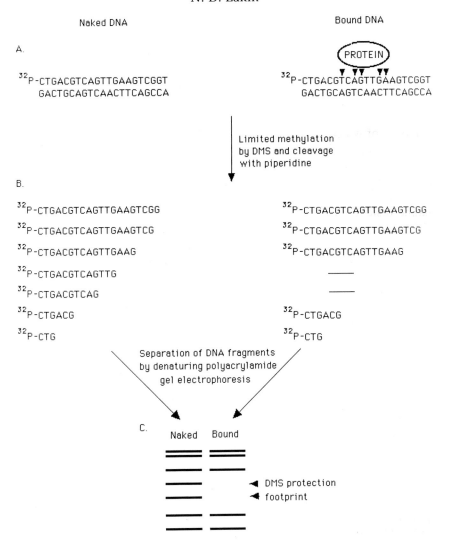

bound DNA). Conversely, DNA contained in the unbound DNA fraction will contain a population of DNA that is methylated randomly on all the guanine residues, due to no protein protection occurring over any of these nucleotides. When this DNA population is subjected to electrophoresis, a complete ladder of guanine residues will be observed. The precise guanine with which the protein interacts can be determined by comparing the guanine ladders produced by the bound and unbound fractions of DNA.

An example of this technique is illustrated in *Figure 4*. As is normally the case, both the sense and anti-sense strands of the DNA have been footprinted to determine the contact points of the protein on both strands of the DNA. In

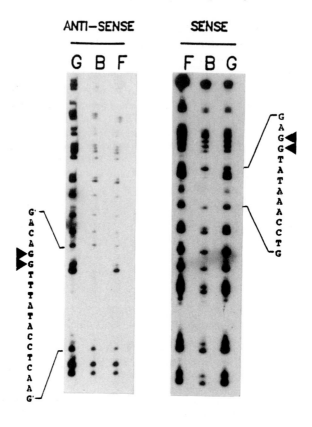

Figure 4. DMS protection footprinting of the SRF binding site of the *X. borealis* skeletal actin gene promoter. A restriction fragment of the *X. borealis* skeletal actin gene promoter that contains an SRF binding site was labelled on either the sense or anti-sense strand of the DNA. This probe was then used in DMS protection footprinting experiments in conjunction with a whole-cell protein extract from *X. laevis* embryos. The bound DNA (track B) and unbound DNA (track F) were separated and isolated as described in *Protocol 4* and analysed by denaturing polyacrylamide electrophoresis after being cleaved at methylated guanine residues by piperidine. Track G represents a Maxam and Gilbert guanine sequencing reaction of the DNA. The sequence of the DNA strand being foot-printed is represented at the side of the DNA ladder. The protection of guanine residues by protein binding is indicated by the arrows. The lower half of the figure indicates (with arrows) those guanine residues within the SRF binding site which are protected.

this example, a short sequence of the *Xenopus borealis* skeletal actin gene promoter is bound by a transcription factor called the serum response factor or SRF (9). Protection from DMS methylation by SRF can clearly be observed over a pair of guanine residues on the sense strand of the DNA, in addition to a pair of guanine residues on the anti-sense strand of the DNA molecule.

The protocol for DMS protection footprinting consists of three steps, namely:

(a) the binding of proteins to DNA and subsequent methylation of guanine residues

(b) separation and isolation of bound and unbound DNA

(c) the cleavage of DNA at methylated guanine residues using piperidine

Protocol 4 describes a procedure for DMS protection footprinting that employs these three steps.

Protocol 4. DMS protection footprinting

A. *The binding of proteins to DNA and subsequent methylation of guanine residues*

1. Assemble a band shift reaction using the appropriate DNA fragment labelled on one strand of the DNA duplex (see Chapter 1, *Protocol 6*). Band-shift reactions should be scaled up by a factor of between 5 and 10 times. This is necessary to enable the subsequent isolation of sufficient bound and unbound DNA fragments for cleavage by piperidine and analysis by electrophoresis.

2. Incubate the reaction mixtures on ice for 30 min before allowing them to stand at room temperature for a further 5 min.

3. Add 1 μl of DMS[a] to each reaction mixture and incubate at room temperature for a further 1 min to partially methylate the DNA (the precise time of incubation with DMS to achieve a satisfactory sequence ladder should be determined empirically).

4. Load the methylated samples directly on to a non-denaturing polyacrylamide gel and electrophorese as described in Chapter 1 (*Protocol 6*).

B. *Separation and isolation of bound and unbound DNA*

1. After electrophoresis, remove one glass plate from the gel and cover the gel on the other plate with cling film. Expose the gel to X-ray film in a light-proof box to visualize the bound (retarded band) and unbound (unretarded band) DNA fractions.

2. Excise the gel fragments containing bound and unbound DNA and cut into fine pieces. Incubate the gel fragments with 300 μl of elution buffer

Protocol 4. *Continued*

   (0.5 M sodium acetate (pH 6.5), 1 mM EDTA (pH 8.0), 0.2% SDS) at 37°C with gentle shaking overnight.

3. Recover and store the elution buffer from each tube, taking care to avoid removing any acrylamide, and rinse the gel pieces in a further 200 μl of elution buffer at 65°C for 5 min. Combine the overnight elution with the rinse from the same sample.

4. To each sample add 10 μg of *E. coli* tRNA and precipitate the nucleic acid by the addition of two volumes of ethanol and leaving at −20°C for at least 1 h.

5. Recover the DNA by centrifugation, wash in 70% ethanol, and dry under a vacuum.

C. *Cleavage of DNA with piperidine*

1. Resuspend each DNA sample in 100 μl of 1 M piperidine.

2. Incubate at 90°C for 3 min to cleave the DNA selectively at methylated guanine residues and allow to cool on ice.

3. Add 1.2 ml of butan-1-ol and vortex until only one phase remains.

4. Recover the DNA by centrifugation in a microcentrifuge for 5 min and discard the supernatant.

5. Dissolve the DNA in 150 μl of 1% SDS and repeat steps 3 and 4.

6. Resuspend DNA in 150 μl of TE (pH 8.0) and extract once with phenol: chloroform (1:1) and once with chloroform; precipitate the DNA by adding sodium acetate (pH 7.0) to 0.3 M and 2 volumes of ethanol.

7. Redissolve the DNA in 5 μl of denaturing loading dye (90% de-ionized formamide, 10 mM EDTA (pH 8.0), 0.01% bromophenol blue, 0.01% xylene cyanol).

8. Load an equal number of c.p.m. of bound and unbound DNA fractions (as determined by their Cerenkov emissions) on to a denaturing polyacrylamide gel and electrophorese as described in *Protocol 3*.

9. Fix the gel in 10% acetic acid and dry under a vacuum before autoradiography.

   a Owing to the toxic nature of DMS, all steps using this reagent must be performed in a fume hood.

---

5. Determination of a transcription factor binding sequence by dimethyl sulphate interference footprinting

The ability of DMS to methylate guanine residues and render the DNA molecule susceptible to cleavage by piperidine at that point has been exploited in another method of DNA footprinting. This technique, known as

DMS interference footprinting, is based on the methylation of a specific guanine residue in the target DNA inhibiting the binding of a transcription factor to that nucleotide. In contrast to DMS protection footprinting, the DNA of interest is partially methylated using DMS prior to the binding of proteins to the DNA. The partially methylated DNA molecules are subsequently mixed with an appropriate protein extract, containing the protein of interest, to allow that protein to complex with the DNA. Following electrophoresis, the band representative of the protein–DNA complex and the band representative of the unbound DNA are isolated from the gel and the DNA cleaved at methylated guanine residues by treatment with piperidine. Clearly, if methylation occurs at a particular guanine residue that is involved in the binding of a protein to the DNA, it will inhibit the protein binding that DNA fragment. Thus, that particular species of DNA will be observed only in DNA which fails to bind the protein. This results in the band of the guanine ladder that is representative of this nucleotide only appearing in the unbound fraction of the DNA when subjected to denaturing gel electrophoresis after cleavage of the DNA molecule with piperidine. Conversely, if a particular guanine residue plays no part in the binding of the protein to the DNA, molecules methylated at this residue will be equally distributed in both the bound and unbound fractions of DNA. Consequently, therefore, if a particular band is absent from the bound fraction of the DNA, then it is indicative of that nucleotide being a contact point of the protein.

Unfortunately, DMS interference footprinting is unable to detect protein-binding sites that contain no guanine residues. This problem can be overcome by methylating all purines to allow the study of interference at adenine and guanine residues simultaneously (for example see reference 10). Alternatively, diethylpyrocarbonate may be employed specifically to modify adenine residues and so render them susceptible to cleavage by piperidine (11).

An example of the DMS interference footprinting technique is illustrated in *Figure 5*. In this example, a footprint is observed that represents a protein binding to equivalent guanine residues contained in the palindromic site B of the negatively acting element of the human immunodeficiency virus promoter (6). As in the case of DMS protection footprinting, both the sense and anti-sense strands of the DNA have been footprinted to establish the contact points of the protein on both strands of the DNA.

The protocol for DMS interference DNA footprinting is similar to that used in DMS protection footprinting, with the exception that the DNA to be footprinted is methylated by DMS before the binding reaction. Thus, this protocol consists of four steps, namely:

(a) limited methylation of guanine residues with DMS

(b) binding of proteins to the methylated DNA

(c) separation and isolation of bound and unbound DNA fractions

(d) cleavage of DNA selectively at methylated guanine residues with piperidine

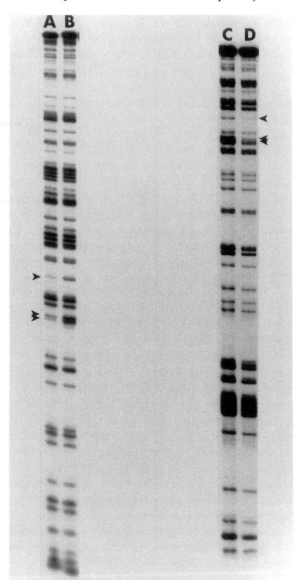

Figure 5. DMS interference footprint of site B of the HIV-1 LTR negative-regulatory element. A restriction fragment containing site B of the HIV-1 LTR negative regulatory element was labelled on either the sense (lanes A and B) or anti-sense strand (lanes C and D) of the DNA duplex. These probes were then used in DMS interference footprinting experiments in conjunction with nuclear protein extracts prepared from Jurkat T-cells. Either unbound DNA (lanes B and C), or DNA recovered from the protein–DNA complex was isolated and the DNA cleaved at methylated guanine residues by piperidine. Arrows indicate the reduction in the intensity of a band that is facilitated by the binding of a protein to a specific sequence of the DNA. Data provided by M. Collins and D. S. Latchman.

Protocol 5 describes a procedure for DMS interference DNA footprinting that employs these four steps.

Protocol 5. DMS interference footprinting

A. *Limited methylation of guanine residues with DMS*

1. Mix 200 μl of 50 mM sodium cacodylate (pH 8.0), 1 mM EDTA with DNA labelled on one strand of the DNA duplex (*Protocols 1* and *2*).
2. Chill to 0°C on ice and add 1 μl of DMS. Mix and incubate at 20°C for 3 min (the optimum time for this reaction should be determined empirically).
3. Add 25 μl of 3 M sodium acetate (pH 7.0) and 0.6 ml of ethanol and incubate at −20°C for at least 1 h to precipitate the DNA.
4. Recover the DNA by centrifugation and dispose of the supernatant by transferring to a waste bottle containing 5 M sodium hydroxide.
5. Wash the pellet with 70% ethanol and dry. Redissolve in 10 μl of TE (pH 8.0).

B. *Binding of proteins to methylated DNA*

1. Assemble a band-shift reaction, using the appropriate partially methylated DNA fragment labelled on one strand of the DNA duplex (for band-shift protocol see Chapter 1, *Protocol 5*). Band-shift reactions should be scaled up by a factor of 5 to 10 times. This is to enable the subsequent isolation of sufficient bound and unbound DNA to be cleaved by piperidine and analysed by denaturing polyacrylamide gel electrophoresis.
2. Incubate on ice for 30 min then load the samples on to a 5% non-denaturing polyacrylamide gel in 0.25 × TBE, and electrophorese as described in Chapter 1 (*Protocol 6*).

C. *Separation and isolation of bound and unbound DNA fractions*

After electrophoresis, isolate the relevant bands of bound and unbound DNA from the gel as described in *Protocol 4 B*.

D. *Cleavage of DNA at methylated guanine residues with piperidine*

Cleave the isolated DNA fragments at methylated guanine residues with piperidine as described in *Protocol 4 C*.

---

6. In vivo DNA footprinting by ligation-mediated polymerase chain reaction

In vitro DNA footprinting has allowed the identification and characterization of a large number of transcription factors. However, the question arises as to whether the observations made *in vitro* are truly representative of the situation in the cell. In an attempt to address this problem, techniques for studying the interaction of proteins with a given DNA sequence *in vivo* have been developed (12–15). However, these techniques require large amounts of starting material and the signal-to-noise ratio is often unacceptable. With the advent of PCR, however, a further *in vivo* footprinting technique has been developed that overcomes these problems by the amplification of a specific sequence ladder from genomic DNA (16).

In vivo DNA footprinting is based on the treatment of cells with DMS which is capable of permeating the cell membrane and causing the limited methylation of DNA sequences *in vivo*. Thus, when DNA is isolated from these cells and treated with piperidine, a population of genomic DNA molecules is generated that are cleaved selectively at methylated guanine residues. By comparing cellular DNA (which is complexed with proteins) with genomic DNA that has been isolated from the cell prior to methylation (naked DNA), a DNA footprint can be observed. The detection of a footprint on a specific DNA sequence amongst the genomic DNA background is achieved by the amplification of a specific DNA sequence by PCR. Unfortunately, however, two defined ends of the DNA are required to amplify a sequence ladder for a specific region of genomic DNA. Although one end of the DNA ladder is fixed by designing one of the PCR primers to be complementary to a specific genomic DNA sequence, the other is determined by the random cleavage of the DNA fragment at methylated guanine residues. In order to overcome this problem, and to create DNA fragments that possess two defined ends, the ligation of a common linker to the variable ends of the DNA is performed before amplification by PCR.

A summary of the methodology of this footprinting technique is illustrated in *Figure 6*. In brief, DNA is first methylated *in vivo* with DMS, isolated, and cleaved at methylated guanine residues with piperidine. DNA is denatured and a primer complementary to a sequence common to all fragments (gene-specific primer) annealed to the DNA. Extension from this primer results in a population of blunt-ended duplex DNA fragments to which a common linker can then be ligated. After ligation of the common linker to the DNA fragments, a second extension is performed, using a primer internal to the first gene-specific primer. This results in a population of DNA fragments that contain two defined ends, namely the linker and second gene-specific primer. Conventional PCR is then performed using the linker and second gene-specific primers to amplify the desired sequence ladder. A third end-labelled primer, whose extending end is 3′ to the second gene specific primer, is then used in a primer extension reaction to allow detection of the specific DNA ladder by autoradiography.

Obviously, if a protein interacts with a specific guanine residue *in vivo*, then these nucleotides will be protected from methylation and subsequent cleavage by piperidine. Therefore, when compared with control naked DNA, the band

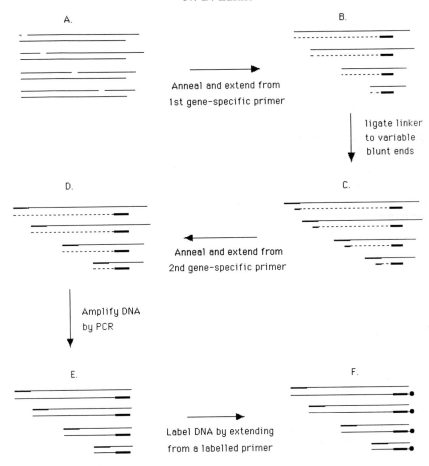

Figure 6. *In vivo* DNA footprinting by ligation-mediated PCR. Genomic DNA is methylated by DMS either *in vivo*, or after isolation from the cell, and subsequently cleaved with piperidine (A). Extension from a gene-specific primer is then performed using the cleaved DNA as a template. This produces duplex DNA fragments with variable blunt ends (B). A defined DNA sequence is then added to the variable ends of the DNA by ligation of a common linker DNA sequence to the blunt ends of the DNA molecules (C). Subsequent extension from a second gene-specific primer yields a population of DNA fragments that have two defined blunt ends (D). These DNA fragments are then amplified by conventional PCR using the second gene-specific primer and a primer complementary to the linker sequence. This results in a population of DNA fragments of varying size, depending on which guanine residue was methylated and cleaved by piperidine (E). These DNA fragments can then be radiolabelled by performing primer extension with a radioactively labelled primer (F) and visualized by autoradiography.

corresponding to this specific nucleotide will be absent from a sequence ladder generated by cleavage of the DNA with piperidine and amplification by ligation-mediated PCR.

An example of how *in vivo* DNA footprinting can be exploited to answer questions about the mechanism whereby transcription factors regulate gene expression is given in reference 16, the study in which this technique was developed. The helix–loop–helix protein MyoD is a transcription factor that binds to the enhancers of muscle-specific genes and activates their transcription during the differentiation of myoblasts into myotubes (for review see reference 17). However, MyoD is expressed both in myotubes, where myocyte-specific genes are active, and in myoblasts, where these genes are silent. Thus, a paradox exists as to how genes activated by MyoD remain silent in myoblasts, despite the expression in this tissue of the proposed myogenic regulatory transcription factor MyoD. The experiments of Mueller and Wold (16) provided an insight into how this may occur. By using *in vivo* DNA footprinting they demonstrated that although MyoD is expressed in myoblasts it does not interact with a MyoD binding site present in the enhancer of the muscle-specific muscle creatine kinase gene. However, this same MyoD-binding site is occupied when the gene is being actively transcribed in myotubes. The authors proposed that additional regulatory mechanisms must exist that restrict the interaction of the MyoD protein with its target site prior to differentiation. Indeed, this would appear to be the case. It has subsequently been demonstrated that the reason MyoD is unable to interact with its target sequences in myoblasts is because a further helix–loop–helix protein, Id, inhibits MyoD binding DNA in the undifferentiated tissue (18).

References

1. Maniatis, T., Fritsch, E. F., and Sambrook, J. (ed.) (1989). *Molecular Cloning: A Laboratory Manual* (second edition), Cold Spring Harbor Press, NY, USA.
2. Krummel, B. (1990). In *PCR Protocols: A Guide To Methods and Applications* (ed. M. Innis, D. H. Gelfand, J. J. Sninsky, and T. J. White), p. 184. Academic Press, London, San Diego.
3. Brown, K. D., Bennett, G., Lee, F., Schweingruber, M. E., and Yanofsky, C. (1978). *J. Mol. Biol.*, **121**, 153.
4. Maniatis, T. and Ptashne, M. (1973). *Nature*, **246**, 133.
5. Galas, J. G. and Schmitz, A. (1978). *Nucleic Acids Res.*, **5**, 3157.
6. Orchard, K., Perkins, N. D., Chapman, C., Harris, J., Emery, V., Goodwin, G., Latchman, D. S., and Collins, M. K. L. (1990). *J. Virol.*, **64**, 3234.
7. Tuillus, T. D. and Dombroski, B. A. (1986). *Proc. Natl. Acad. Sci. USA*, **83**, 5469.
8. Maxam, A. and Gilbert, W. (1980). *Methods Enzymol.*, **65**, 499.
9. Lakin, N. D., Boardmen, M., and Woodland, H. R. (1992). Unpublished observations.

10. Ares, M. Jr., Chung, J.-S., Giglio, L., and Weiner, A. M. (1987). *Genes Development*, **1**, 808.
11. Sturm, R. A., Das, G., and Herr, W. (1988). *Genes Dev.*, **2**, 1582.
12. Church, G. M. and Gilbert, W. (1984). *Proc. Natl. Acad. Sci. USA*, **81**, 1991.
13. Huibregtse, J. M. and Engelke, D. R. (1986). *Gene*, **44**, 151.
14. Jackson, P. D. and Felsenfeld, G. (1985). *Proc. Natl. Acad. Sci. USA*, **82**, 2296.
15. Mueller, P. R., Salser, S. J., and Wold, B. (1988). *Genes Dev.*, **2**, 412.
16. Mueller, P. R. and Wold, B. (1989). *Science*, **246**, 780.
17. Olson, E. N. (1990). *Genes Dev.*, **4**, 1454.
18. Benezra, R., Davis, R. L., Lockshon, D., Turner, D. L., and Weintraub, H. (1990). *Cell*, **61**, 49.

3

Biochemical characterization of transcription factors

AUSTIN J. COONEY, SOPHIA Y. TSAI, and
MING-JER TSAI

1. Introduction

Once a *cis*-element and transacting factor have been determined to be important for transcription of a gene, gel retardation, DNase I footprinting, methylation interference analyses, and *in vitro* transcription assays are generally employed for characterization of a factor during purification. The majority of transcription factors are present in very low quantities within a cell, making purification long and arduous, and requiring large quantities of starting material. Thus the initial characterization of the factor is undertaken with partially purified material. This permits elucidation of vital information about the factor, such as the molecular weight, subunit composition, and interactions, if any, with non-DNA-binding factors, before complete purification of the factor (1). This information is mostly derived from analysis of data from a combination of chromatographic and electrophoretic techniques.

2. Determination of the molecular weight of the native factor

Molecular weight determination under native conditions allows the researcher to estimate the total size of the factor/complex binding to an element and may aid in identifying whether the binding species is a monomer, dimer, or higher complex. Separation of factors, on the basis of size, under native conditions has the advantage of allowing direct assay of DNA-binding and transcriptional activities, to assign the factor's molecular weight relatively easily with samples where purity is not very high. In order to determine accurately the molecular weight of the factor under native conditions, the factor must be fairly globular in conformation. Transcription factors exist as an arrangement of globular domains and thus lend themselves to this type of analysis.

2.1 Gel filtration chromatography

Gel filtration chromatography separates proteins on the basis of their size, where size refers to the physical dimensions or Stoke's radius, rather than the molecular weight, which is related to mass. However, in the case of globular proteins that have a roughly spherical tertiary structure, the Stoke's radius is proportional to the molecular weight. Thus a semi-logarithmic standard curve can be produced plotting K_{av} versus log molecular weight, where K_{av} is the partition coefficient (see *Figure 1*). The partition coefficient is defined by the equation:

$$K_{av} = \frac{V_e - V_0}{V_t - V_0}$$

where V_e = elution volume of the protein, V_0 = void volume, and V_t = total volume.

Thus the molecular weight of an unknown factor can be determined by linear regression from such a standard curve constructed from the chromatography of standard proteins of known molecular weight. The gel filtration matrix consists of porous beads that molecularly sieve proteins from the largest to the smallest. Gel filtration achieves two purposes: it permits determination of the molecular weight of a native factor binding to a response element (2, 3) and it also achieves an additional step in the purification of the transcription factor (4, 5). Gel filtration columns can be calibrated with gel filtration protein standards of known molecular size (Bio-Rad Laboratories, Pharmacia LKB Biotechnology, or Sigma Chemical). Identification of the elution volume of the DNA-binding or transcriptional activities allows determination of the molecular weight of the factor by reference to the standard curve. Selection of the appropriate gel filtration matrix is important. Initially, for an uncharacterized transcription factor whose molecular weight is unknown, a matrix with a broad fractionation range is appropriate. However, once the molecular weight of the factor has been estimated, the most accurate determination can be achieved with a matrix from which the factor will elute midway in the fractionation range, where the plot of K_{av} vs. log molecular weight is linear (see *Figure 1*). Sephacryl S-300 (Pharmacia) or Bio-Gel A-1.5M (Bio-Rad) have excellent broad fractionation ranges from 10^4 to 1.5×10^6 relative molecular mass (M_r), and are useful in the initial determinations of molecular weights. A number of different matrices, with different fractionation ranges and characteristics, are produced by Pharmacia LKB and Bio-Rad (see *Table 1*). A potential, but infrequent, problem with gel filtration matrices is non-specific interaction between a transcription factor and the gel matrix, leading to retardation of its elution or the opposite effect. This has been observed for some steroid receptors when agarose A-15M or A-1.5M (Bio-Rad) is used; where this phenomenon is suspected, i.e. $K_{av} > 1$, change the matrix from Sepharose to acrylamide or vice versa. However, the molecular weight of many native factors have been determined

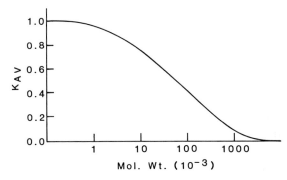

Figure 1. The sigmoidal dependence of K_{av}, the partition coefficient, on the logarithm of the molecular weight (mol. wt.).

Table 1. Fractionation ranges of some commercially available gel filtration matrices

Matrix Type[a]	Fractionation range[b]
Sephadex G-10	$\leqslant 7 \times 10^2$
Sephadex G-15	$\leqslant 1.5 \times 10^3$
Bio-Gel P-2	$1 \times 10^2 - 1.8 \times 10^3$
Bio-Gel P-4	$8 \times 10^2 - 4 \times 10^3$
Sephadex G-25	$1 \times 10^3 - 5 \times 10^3$
Bio-Gel P-6	$1 \times 10^3 - 6 \times 10^3$
Bio-Gel P-10	$1.5 \times 10^3 - 2 \times 10^4$
Sephadex G-50	$1.5 \times 10^3 - 3 \times 10^4$
Bio-Gel P-30	$2.5 \times 10^3 - 4 \times 10^4$
Bio-Gel P-60	$3 \times 10^3 - 6 \times 10^4$
Sephadex G-75	$3 \times 10^3 - 8 \times 10^4$
Sephacryl S-100	$1 \times 10^3 - 1 \times 10^5$
Bio-Gel P-100	$5 \times 10^3 - 1 \times 10^5$
Sephadex G-100	$4 \times 10^3 - 1.5 \times 10^5$
Sephacryl S-200	$5 \times 10^3 - 2.5 \times 10^5$
Sephadex G-150	$5 \times 10^3 - 3 \times 10^5$
Sephadex G-200	$5 \times 10^3 - 6 \times 10^5$
Bio-Gel A-0.5M	$1 \times 10^4 - 5 \times 10^5$
Sephacryl S-300	$1 \times 10^4 - 1.5 \times 10^6$
Bio-Gel A-1.5M	$1 \times 10^4 - 1.5 \times 10^6$
Sepharose CL-6B	$1 \times 10^4 - 4 \times 10^6$
Bio-Gel A-5M	$1 \times 10^4 - 5 \times 10^6$
Bio-Gel A-15M	$4 \times 10^4 - 1.5 \times 10^7$
Sepharose CL-4B	$6 \times 10^4 - 2 \times 10^7$
Sepharose CL-2B	$7 \times 10^4 - 4 \times 10^7$
Bio-Gel A-50M	$1 \times 10^5 - 5 \times 10^7$
Bio-Gel A-150M	$1 \times 10^6 - 1.5 \times 10^8$

[a] Sephadex, Sephacryl, and Sepharose are supplied by Pharmacia LKB Biotechnology and Bio-Gel is supplied by Bio-Rad.
[b] The fractionation range denotes the molecular sizes of proteins that elute in a volume equal to the bed volume to proteins that elute in the void volume.

by gel filtration (1). Pharmacia also produces pre-poured gel filtration columns (Superose) for use with their FPLC system, which can be used for all the purposes described here.

Protocol 1. Molecular weight determination by gel filtration

1. Swell the dry gel matrix in the appropriate buffer or according to the manufacturer's directions. Allow the gel matrix to settle and aspirate the fines. Alternatively, if the gel matrix comes pre-swollen, wash the matrix in a sintered-glass funnel to remove the preservatives and to equilibrate it in the gel filtration buffer. De-gas the slurry.

2. Erect the column vertically in a 4°C cold room. Optimum resolution of proteins is best achieved with columns greater than 50 cm in length with a matrix diameter of 1 cm.

3. Fill the outlet tube with buffer and clamp it. Then add some buffer to the column gel matrix suspension and pour slowly to the desired bed height. Great care should be taken in pouring the column; it should be poured in a single motion to produce a homogeneous column matrix with no air bubbles.

4. Layer 1 cm of buffer carefully on top of the column bed, after the gel matrix has settled, without disturbing the matrix. Then connect the gel filtration buffer reservoir to the column. Open the column outlet and wash the gel matrix with 2 to 3 column-volumes of the gel filtration buffer.

5. Determine the void volume (V_0) and the total volume (V_t) of the column by filtration of dextran blue and potassium bichromate, respectively.

6. Carefully allow the buffer in the column to drop to the level of the bed and close the outlet. Disconnect the reservoir and gently load the protein standards (10–20 μg each in a volume equivalent to 1–2% of the total column volume.) Open the outlet and let the protein standards enter the column, then close the outlet. Carefully pipette a 1 cm layer of gel filtration buffer on to the column bed and reconnect the buffer reservoir.

7. Open the outlet and commence elution of the protein standards at a flow rate of 8 ml/h. (The flow rate varies from column matrix to matrix and is determined by multiplying the linear flow rate (supplied by the manufacturer) by the cross-sectional area of the column in square centimetres to yield the flow rate in millilitres per hour.) Collect fractions equal to 1–2% of the total column volume using an automated fraction collector with an on-line UV detector. Alternatively, aliquots of each fraction can be taken to determine the protein peaks of the standards by measuring the absorbance at 280 nm (A_{280}).

8. Electrophorese aliquots of each fraction containing a protein peak on a

denaturing 10% polyacrylamide-SDS gel, with a stacking gel using the discontinuous buffer system of Laemmli (6), to determine their molecular weight. This calibrates the column and allows the preparation of a standard curve to predict the size of a polypeptide which elutes in a particular fraction from the column.

9. Load the protein sample at a concentration of 0.2–0.5 mg/ml in a volume equal to 1–2% of the column volume. The loading and elution of the transcription factor preparations are the same as for the protein standards (see steps 5 and 6).

10. Measure the A_{280} of aliquots from each fraction. Take 5–10 μl aliquots from alternate fractions to assay for specific DNA binding, by use of gel retardation, DNase I footprinting assays (see Chapters 1 and 2), and/or transcriptional activity (see Appendix).

2.2 Glycerol gradient centrifugation

An alternative method can be used to determine and to confirm the native size determined by gel filtration and to eliminate possibilities of non-specific interactions between the gel matrix and the factor. This method uses density gradient sedimentation developed by Martin and Ames (7). The formation of a glycerol gradient and sedimentation of the applied sample by ultracentrifugation permits the fractionation of partially purified factors on the basis of their sedimentation coefficients, which approximates to their molecular weight (3). Determination of molecular weight by sedimentation velocity has the advantage over gel chromatography of being a primary technique which permits the direct determination of hydrodynamic parameters, such as the sedimentation coefficient (s), from first principles. See the reference of Siegel and Monty for a detailed discussion of the derivation and use of the equations (8).

However, protein standards of known s can be used to determine molecular weights rapidly and easily by glycerol gradient centrifugation. The use of glycerol or sucrose to form the gradient has the advantage, because of their viscosity and density, that essentially linear migration of the majority of biological macromolecular results. Thus the ratio (R) of the distances sedimented by any two substances will always be constant. Given a standard protein of known s, the s of an unknown factor can be calculated from the equation:

$$R = \frac{\text{distance travelled from the meniscus by unknown}}{\text{distance travelled from the meniscus by standard}}$$

As the rate of movement of any macromolecule is nearly constant and assuming they have the same partial specific volumes then:

$$R = \frac{s_{20,w} \text{ of unknown } (s1)}{s_{20,w} \text{ of standard } (s2)}$$

where $s_{20,w}$ is the sedimentation coefficient at 20°C in water

An approximate value for the molecular weight can be determined by:

$$\frac{s1}{s2} = \left(\frac{MW1}{MW2}\right)^{\frac{2}{3}}$$

where MW = molecular weight.

The glycerol gradient can be calibrated by sedimenting protein standards of known molecular weight (Bio-Rad) whose position in the gradient can be determined by electrophoresing aliquots from alternate fractions on an SDS-polyacrylamide gel (6). Identification of the position in the glycerol gradient to which the transcription factor sediments is determined by assaying fractions of the gradient for specific DNA-binding and/or transcriptional activity. Measurement of the distance travelled by the unknown factor and standards permits calculation of the factor's molecular weight. This technique has been commonly used to determine the native molecular weight of transcription factors (1, 3, 9).

Protocol 2. Determination of molecular weights by glycerol gradient centrifugation

1. Pour 4.4 ml linear 7–23% (w/v) glycerol gradients, in 20 mM Hepes (pH 7.9), 100 mM KCl, 4 mM dithiothreitol (DTT) and 0.2 mM EDTA in SW50.1 tubes using a Beckman Density Gradient Former and a mixing chamber.

2. Layer a 200 μl aliquot of partially purified or affinity purified transcription factor carefully on to the pre-formed glycerol gradients. This sample should contain 15 IU of SP6 polymerase (Promega) as an internal control.

3. Layer 10 to 20 μg of molecular weight standards (Bio-Rad) carefully on to a balance gradient to construct a standard curve to calibrate the gradient.

4. Centrifuge the samples at 250 000 × *g* (45 000 r.p.m.) in an SW50.1 rotor (Beckman Instruments, Inc.) for 16 h at 4°C.

5. Collect fractions of 150–200 μl from each gradient from the top to the bottom with an Auto Densi-Flow (Buchler) connected to a fraction collector, or puncture the bottom of the tube and collect fractions from the bottom up.

6. Assay each fraction for SP6 polymerase activity as per the manufacturer's instructions. It has a molecular weight of 98 kDa.

7. Electrophorese an aliquot of each fraction from the standard curve glycerol gradient under denaturing conditions in a 10% polyacrylamide-SDS gel (6). The position of the standards are detected by silver or Coomassie blue staining (10) to determine in which fractions they are located, to calibrate the glycerol gradients.

8. Assay 5–10 μl aliquots of each fraction for specific DNA binding, by gel retardation or DNase I footprinting, and/or transcriptional activity to identify the fraction in which the transcription factor of interest sediments. This allows determination of the molecular weight by comparison with the standard curve.

2.3 Non-denaturing gradient gel electrophoresis

Non-denaturing gradient gel electrophoresis can be used to determine the native molecular weight of a transcription factor (11, 12). Gradients of polyacrylamide yield greater resolution of a protein mixture compared with a single polyacrylamide concentration and are thus more accurate for molecular weight determinations of native proteins. If the factor has been purified to homogeneity, it can be identified by silver staining (10), and if it has only been partially purified and an antibody is available, it can be identified by Western analysis (11, 13). Here we will describe non-denaturing gradient electrophoresis and Western analysis. Although we describe the traditional use of radio-iodinated protein A for the detection of proteins in Western analysis, many excellent non-radioactive kits are available from several manufacturers, with accompanying protocols, which allow rapid sensitive, safe, and reproducible analysis of Western blots. Many transcription factors in their native form are associated with other factors in relatively large complexes or as dimers, thus to facilitate the electrophoretic transfer of the factor, 20% of the N,N'-methylenebisacrylamide cross-linker can be replaced with the reversible cross-linker diallytartardiamide (DATD): Incorporation of DATD allows the polyacrylamide gel to be permeabilized by brief incubation with 5 mM periodic acid. A drawback of non-denaturing gradient electrophoresis is that it yields the best results with highly purified transcription factor preparations. Crude nuclear or cytosolic extracts tend to form insoluble aggregates which do not enter the gel and remain in the well and also tend to produce ill-defined smeared bands on the Western immunoblot.

Protocol 3. Molecular weight determination by non-denaturing gradient electrophoresis

1. Cast a 3–25% gradient polyacrylamide gel (acrylamide–bisacrylamide–DATD, 79: 0.8:0.2 by wt) in 0.5 × TBE (50 mM Tris, 50 mM boric acid, 1 mM EDTA) with a Bio-Rad gradient gel casting chamber. A 12 ml/min flow rate was maintained by a Gilson peristaltic pump (Gilson, Oberlin OH).

2. Pre-run the gel at 300 V for 3 h to overnight in 0.5 × TBE at 4°C.

3. Incubate the samples at room temperature for 10–15 minutes. At this stage, DNA or ligands for particular experiments can be added.

Protocol 3. *Continued*

4. Load 200 ng to 1 μg of transcription factor preparation and pre-stained molecular weight markers (Bio-Rad). Run at 300 V for 5 h at 4°C.

5. Rinse the gel in distilled H_2O three times.

6. Incubate the gel in 5 mM periodic acid for 15 min at 30°C.

7. Repeat step 5.

8. Incubate in transfer buffer (49 mM Tris, 39 mM glycine, 0.04% SDS (pH 9.2)) for 10 min.

9. Place the gel on a pre-cut, pre-wetted nitrocellulose or nylon membranes, such as Zeta-Probe or Immobilon P, and sandwich between pre-wetted Whatman No. 3 paper, three sheets on each side.

10. Place the sandwich in trans-blot apparatus and run at 20 V for 1 h in transfer buffer.

11. Disassemble the blot and block the membrane for 1 h at room temperature with blotto (2–3% non-fat dry milk (w/v), 10 mM Tris–HCl (pH 8) and 150 mM NaCl).

12. Decant the blotto and add the appropriate amount of antibody in 8–10 ml blotto. Incubate for 3–4 h at room temperature with gentle shaking.

13. Decant the antibody solution and wash the membrane with 40 ml Western buffer (10 mM Tris–HCl (pH 8.0) and 150 mM NaCl) for 5 min with shaking. Decant and repeat twice.

14. If a monoclonal antibody was used, add the secondary antibody (rabbit anti-mouse IgG, commercially available) in 10 ml blotto and incubate at room temperature for 90 min with gentle shaking. If a polyclonal antibody was used, go straight to step 16.

15. Repeat step 13.

16. Add 2 μl of ^{125}I labelled Protein A (1.25 mCi-ICN) to 10 ml blotto. Incubate at room temperature with gentle shaking for 90 min.

17. Decant and dispose of the radioactive blotto carefully. Wash the membrane with 40 ml Western buffer for 10 min at room temperature. Decant and repeat twice.

18. Dry the membrane and develop the signal by autoradiography.

---

3. Determination of the molecular weight of the denatured factor

Analysis of a factor under native conditions only permits estimation of the molecular weight of the total complex binding to an element. Electrophoresis under denaturing conditions (SDS-polyacrylamide gel electrophoresis (SDS-

PAGE)) allows accurate determination of the molecular weight of the individual transcription factors involved in that complex (3, 6, 14, 15). Comparison of the results will also yield invaluable insights into the factors binding to an element. If the factor binds as a dimer, then the native molecular weight will be approximately double that of the denatured factor. Also, if the complex is composed of a heterodimer, it may be possible to identify two species by SDS-PAGE if they have different molecular weights. Determination of the molecular weight of a factor under denaturing conditions requires the identification of a specific band on an SDS-polyacrylamide gel which corresponds to the binding or transcriptional activity being studied. The two methods outlined here differ conceptually as to whether the factor is identified before or after running the SDS-polyacrylamide gel. Ultraviolet (UV) cross-linking to DNA identifies the binding activity prior to running the SDS gel, while renaturation and assay for binding or transcriptional activity identifies the factor after running the gel. Both of these procedures have been employed successfully to identify the molecular size of many transcription factors early in the purification procedure.

3.1 UV cross-linking to DNA

Irradiation of DNA with UV light produces purine and pyrimidine free radicals that are chemically reactive and can form covalent bonds such as thymidine dimers. This reactive property of UV-irradiated DNA can be used to link transcription factors covalently to their response elements. When a protein–nucleic acid complex is irradiated with UV light, it causes the formation of covalent bonds between pyrimidines and certain amino acid residues in the DNA-binding domain in close proximity to the DNA. Thus a transcription factor can be selectively labelled indirectly by UV cross-linking as a consequence of its specific binding to a DNA sequence (16). Labelling the transcription factor in this fashion allows the easy and rapid determination of its molecular weight, under denaturing conditions in an SDS-polyacrylamide gel, even in crude extracts. We describe two techniques here, one involves incorporation of deoxybromouridine into the DNA prior to cross-linking, the other utilizes UV cross-linking to an unsubstituted oligonucleotide and gel retardation assay.

3.2 UV cross-linking with a bromodeoxyuridine-substituted DNA

Halogenated analogs of thymidine (e.g. bromodeoxyuridine) are significantly more sensitive to UV-induced cross-linking because replacement of the thymidine methyl group with a bromine atom creates a molecule more susceptible to UV-light-induced free-radical formation (17). Bromodeoxyuridine can be functionally utilized in place of thymidine by many enzymes including the Klenow fragment, which is used to prepare the substituted probe. An additional advantage of using bromodeoxyuridine is that a UV light of longer wavelength can be used for cross-linking than can be used with unsubstituted DNA. Longer wavelength UV light is less damaging to DNA because it causes fewer nicks in the phosphate backbone. After cross-linking of the factor and DNA, the probe is digested to remove the majority of the probe not protected by and covalently linked to the factor. This is a precautionary measure since UV light may also induce the formation of thymidine dimers in the DNA, leading to electrophoretic anomalies. Thus the removal of excess DNA will minimize this problem.

Protocol 4. Cross-linking with a bromodeoxyuridine-substituted DNA

1. Subclone a DNA fragment or oligonucleotide containing the specific response element for the transcription factor into M13.

2. Mix 5 µg of the recombinant single-stranded M13 clone containing the binding site, with an equimolar amount of the 17 nucleotide M13 universal primer. Add 10 µl of 10 × Klenow buffer (50 mM Tris–HCl (pH 7.5), 10 mM MgCl$_2$, 1 mM DTT, 50 mg/ml bovine serum albumin (BSA)).

3. Heat at 95°C for 5 min and cool very slowly at room temperature, this should take several hours.

4. Make the following additions to the annealed mixture:
   - 50 µl [α^{32}P]dCTP (3000 Ci/mmol)
   - 3.5 µl of dNTP mix (containing 2.5 mM each of dATP, dGTP, 5-bromo-2'-deoxyuridine trisphosphate (Pharmacia), and 0.25 mM dCTP)
   - 1.75 µl 0.1 M DTT
   - 7.5 µl 10 × Klenow buffer
   - 7 µl H$_2$O
   - 5 µl Klenow fragment enzyme (25 IU, Promega)

5. Incubate at 16°C for 90 min.

6. Heat the reaction to 68°C for 10 min to inactivate the Klenow enzyme.

7. Digest the double-stranded M13 recombinant vector with an excess of enzymes, under appropriate reaction conditions, that will liberate the inserted response element.

8. Extract the reaction once with a mixture of phenol:chloroform:isoamyl-alcohol (25:24:1, by vol.) Microcentrifuge for 5 min at 11 000 × g and aspirate the aqueous phase.

9. Precipitate the DNA with 0.3 M ammonium acetate and 2 volumes of 100% ethanol, incubate in a dry ice/ethanol bath for 10 min and micro-centrifuge for 15 min at 11 000 × g. Wash the pellet with 0.5 ml 70% ethanol. Dry the pellet and resuspend in 20 µl TE buffer (pH 8) and add 2 µl 10 × loading buffer (0.25% bromophenol blue, 0.25% xylene cyanol, and 25% Ficoll (type 400) in H$_2$O).

10. Load the DNA on to an agarose gel and electrophorese. Isolate the required fragment by your favourite technique (or Geneclean (Bio 101) or NA45 DEAE membrane (Schleicher and Schuell), according to the manufacturers' protocols).

11. Dilute and count an aliquot of the probe using a scintillation counter and determine the DNA concentration by measuring the A_{260}. Then determine the specific activity of the probe.

12. Set up a number of binding reactions. Incubate at room temperature under conditions previously determined (generally those used for gel retardation assays), for 15–30 min. The reactions should contain 10^5 c.p.m. of labelled probe, the extract or fraction containing the transcription factor, transcription buffer (20 mM Hepes (pH 7.9), 100 mM KCl, 20% glycerol, 4 mM DTT, 0.2 mM EDTA) and 10–20 μg of non-specific competitor, e.g. poly(dI-dC)·poly (dI-dC); bring the volume up to 50 μl with H_2O. Check for successful binding by running a gel retardation assay (Chapter 1).

13. Cover the top of the tube with UV-transparent cling film.

14. Irradiate the binding reaction from a distance of 5 cm (see *Figure 2*) with a UV transilluminator emitting at 305–310 nm, with a maximum intensity of 7000 μW/cm^2 (e.g. a Fotodyne UV lamp) for 5–60 min. The effect of UV irradiation can also be checked by running a gel retardation assay.

15. Add 1 μl 0.5 M $CaCl_2$, 4 mg DNase I (Worthington Diagnostics) and 1 IU micrococcal nuclease (Worthington Diagnostics) to each reaction and incubate at 37°C for 30 min.

16. Add an equal volume of 2 × SDS sample buffer (0.125 M Tris–HCl (pH 6.8), 6% SDS, 10% β-mercaptoethanol, 20% glycerol, 0.025% bromophenol Blue) to each binding reaction and boil for 5 min.

17. Load the samples on a 10% polyacrylamide-SDS gel with a stacking gel, include a marker lane containing ^{14}C-labelled protein standards (BRL, Life Technologies, Inc.).

18. Electrophorese for 2–3.5 h at 35 mA (6). Remove the portion of the gel at the dye front, which contains the majority of the radioactivity in the digested nucleic acids.

19. Fix the gel with 7.5% acetic acid and 50% methanol for 60 min. Then impregnate the gel with a fluor, to enhance detection (e.g. Enhance, NEN) according to the manufacturer's instructions.

20. Dry the gel for 1 h at 60°C and then 1 h at 80°C. Then autoradiograph with Kodak XAR-5 X-ray film and a Dupont Cronex Lightening Plus intensifying screen. Expose the film at −70°C for 1–4 days to visualize the cross-linked protein.

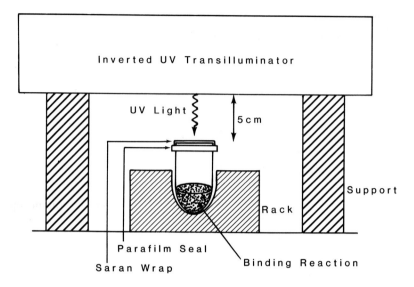

Figure 2. Experimental set-up for UV irradiation of binding reactions.

3.3 UV cross-linking of non-substituted probes

DNA can be cross-linked to a bound transcription factor in the absence of substitution with bromodeoxyuridine, however, at lower efficiency (1). In this case cross-linking occurs directly between the DNA bases and the poly-peptide. In the protocol we describe here, an end-labelled oligonucleotide is UV cross-linked *in situ* as the final step in a gel retardation assay. A short oligonucleotide should be chosen, to minimize the effects of UV light on the electrophoretic mobility of the DNA because the probe will not be digested with nucleases. The advantage of this procedure is that labelling an oligo-nucleotide obviates the time-consuming subcloning and probe preparation in M13. The gel retardation assay is a very useful analytical tool for identify-ing and studying the binding of specific factors. It can also be used as a preparative procedure. Gel retardation both identifies and separates a factor from the milieu of crude extracts or partially purified fractions. Use is made of these dual properties to determine the molecular weight of a factor which binds to a specific element. In order to increase the success of cross-linking a transcription factor to DNA, the binding sequence must be A:T rich, as thymidine is the most reactive of the bases. Also certain response elements bind multiple proteins and the gel retardation assay allows individual factor–DNA complexes to be visualized and isolated, and then analysed by SDS-PAGE.

Protocol 5. UV cross-linking of non-substituted DNA

1. To end-label the oligonucleotide, make the following additions:
   - 100 ng of double-stranded oligonucleotide
   - 100 µCi each of $[\alpha\text{-}^{32}P]dCTP$, $[\alpha\text{-}^{32}P]TTP$, $[\alpha\text{-}^{32}P]dATP$, and $[\alpha\text{-}^{32}P]dGTP$
   - 5 µl Klenow fragment (25 IU)
   - 5 µl 10 × Klenow buffer
   - H_2O to a final volume of 50 µl

   Incubate at room temperature for 60 min.

2. Stop the reaction by adding 4 µl 0.5 M EDTA, bring the volume up to 100 µl with TE buffer (pH 7.5) and extract with an equal volume of phenol:chloroform. Aspirate the aqueous phase.

3. Add one-tenth of a volume of sodium acetate (pH 5.2) and 3 volumes of ethanol. Incubate in a dry ice/ethanol bath for 30 min.

4. Microcentrifuge at 11 000 × g at 4°C for 40 min and discard the radio-active supernatant carefully.

5. Resuspend the pellet in 100 µl 1 M ammonium acetate, vortex vigorously, add 250 µl ethanol and incubate in a dry ice/ethanol bath for 30 min.

6. Microcentrifuge for 40 min at 11 000 × g at 4°C and aspirate the super-natant.

7. Repeat steps 5 and 6.

8. Wash the pellet with 0.5 ml 80% ethanol, dry the pellet and resuspend it in 50 µl TE buffer (pH 8).

9. Determine the specific activity of the probe (see *Protocol 4*, step 11).

10. Incubate the partially purified transcription factor with the probe under binding conditions previously determined (see *Protocol 4*, step 12).

11. Cross-link the probe and transcription factor. The UV light wavelength used should be 254 nm. The duration of irradiation can be varied from 5 min to over 3 h, as this reaction is less efficient.

12. Electrophorese each reaction in a native 5% polyacrylamide gel to separate the protein–DNA complexes from the unbound probe.

13. Autoradiograph the gel for 1–3 h at 4°C with Kodak XAR-5 X-ray film.

14. Excise the retarded protein–DNA complexes.

15. Layer the gel slices on to a 10% polyacrylamide-SDS protein gel with a stacking gel. Load a lane of ^{14}C-labelled molecular weight standards.

16. Electrophorese and autoradiograph the gel as per *Protocol 4* (see steps 17–20).

3.4 Renaturation of transcription factors from SDS-polyacrylamide gels

Fractions containing partially purified transcription factors can be separated on the basis of size, with high resolution, by SDS-PAGE. Proteins eluted and renatured from the gel can be assayed for specific DNA-binding and/or transcriptional activity. This technique allows accurate determination of the molecular weight of individual factors that bind to an element. However, this is dependent on the factor being able to bind to DNA as a monomer or a homodimer. If the active form is a heterodimer, specific DNA-binding or transcriptional activity may not be detected by this procedure. The most critical parameter of this procedure is the ability of a transcription factor to renature into its active form. Many transcription factors are composed of domains with distinct structural conformations, which aids the process of renaturation. However, the likelihood of successful renaturation increases with decreased size. The recovery of active transcription factor after this procedure is very low. On average, a 5% recovery is assumed (this will vary from factor to factor and also with size) to calculate the amount of the factor that has to be loaded on the gel to yield enough active factor after renaturation to be detectable by *in vitro* transcription or gel mobility shift assays. The method of Hager and Burgess (18), with a few modifications, is recommended and has been successfully employed to determine the molecular size of transcription factors (19).

Protocol 6. Renaturation from SDS-polyacrylamide gels

1. Add one-fifth volume of 5 × SDS loading buffer (0.25 M Tris–HCl (pH 6.8), 15% SDS, 50% glycerol, 0.025% bromophenol blue; before boiling add to each sample β-mercaptoethanol to 5%) to an aliquot of up to 100 μl of partially purified or affinity-purified transcription factor and boil for 2 min.

2. Load the supernatant on to a denaturing SDS-polyacrylamide gel (7.5–10%, depending on protein size) with a stacking gel (6).

3. Electrophorese for 3–5 h at 30 mA (6).

4. Remove the molecular weight markers and a portion of the sample lane and silver stain (10).

5. The unstained sample lane is cut into 5–10 mm strips from the bottom to the top of the gel.

6. Place each gel in an individual small dialysis bag (M_r cut-off 10 000) containing 0.5 ml of SDS-gel running buffer and 100 μg/ml BSA (nuclease free, supplied by Boehringer Mannheim Biochemicals). Place the dialysis bags perpendicular to the electric current in a large horizontal gel trough, filled with SDS-gel running buffer. The current is applied at 50 mA for 3 h to elute the proteins from the gel.

7. Recover the buffer containing the electroeluted proteins from the dialysis bags and precipitate by the addition of acetone (4:1, v:v) in 15 ml corex tubes and incubate at −70 °C for 30 min or −20 °C overnight. The precipitate is centrifuged at 10 000 × g for 30 min at 4 °C. This step concentrates the protein and removes the SDS.

8. Rinse the pellet once with 80% acetone and 20% transcription buffer and dry.

9. Dissolve the pellets in 100 µl of 6 M guanidine-hydrochloride in transcription buffer and incubate at room temperature for 20 min. This treatment denatures the polypeptides.

10. Place the resulting solutions in individual small dialysis bags and dialyse against transcription buffer for 16 h at 4 °C, with one change of buffer during that period of time.

11. Assay 5–10 µl of the dialysed samples for specific DNA binding by gel mobility shift assay or DNase I footprinting (see Chapters 1 and 2) and/ or transcriptional activity (see Appendix).

---

4. Analysis of the monomer/dimer structure of the DNA-binding forms of a transcription factor

Another characteristic of transcription factors requiring examination is the subunit composition of the active DNA-binding form. Many transcription factors exist as monomers, dimers, or even trimers as their active DNA-binding forms. At this stage of characterization of a transcription factor there should be substantial evidence to indicate its subunit composition. One indicator is the sequence of the DNA response element itself. Symmetry within the sequence of the response element, whether it is palindromic or directly repeated, suggests that the DNA-binding form of the transcription factor may be a dimer or a higher order structure. Symmetric response elements are the binding elements for many groups of transcription factors, including helix–loop–helix factors and the steroid/thyroid hormone receptor superfamily. Both families bind to their cognate response elements as dimers. DNase I protection and methylation interference analysis will provide evidence as to whether a factor protects and contacts the repeated sequence motif (see Chapter 2). The response element will then indicate the subunit structure of the transcription factor.

Comparison of the molecular weights of a transcription factor determined under both native and denaturing conditions will provide more direct evidence of the subunit structure of the DNA-binding form. If a factor binds as a monomer, its molecular weight determined under native and denaturing conditions will be the same. If the factor binds as a dimer, its native molecular weight will be approximately double the molecular weight determined under denaturing conditions, and so on for higher order structures. If the native molecular weight is not an exact whole-number multiple of the denatured molecular weight, this may indicate that the DNA-binding form is a hetero-dimer of two subunits with different molecular sizes. Heterodimers have been observed for many transcription factors. With this preliminary information, more direct experiments are required to determine the subunit structure of the transcription factors. These experiments involve analysis by gel retarda-tion and chemical cross-linking. In addition, co-operative DNA binding can be observed between dimers bound to adjacent response elements. This co-operativity can be analysed by gel mobility shift assays.

4.1 Analysis by gel retardation

The ternary structure of transcription factors can be analysed indirectly by gel retardation. The use of double-stranded oligonucleotides as probes allows easy and rapid analysis of factor-binding requirements by incorporation of base mutations into the sequence during synthesis. Methylation interference will often highlight the symmetric nature of the purine contact points. These nucleotides are good points to start mutational analysis, by changing them to pyrimidines and analysing their effect on factor binding by gel retardation. In this fashion, mutation of certain residues of a half site of the symmetric response element may lead to loss of binding activity. This suggests that the factor binds as a dimer, requiring binding of both subunits.

To provide absolute proof that the active DNA binding form is a dimer, the factor needs to be cloned (Chapters 4–6). The cloned gene permits genetic manipulations, including deletion analysis and formation of fusion proteins, which allows analysis of structure–function relationships, such as dimer for-mation (see Chapter 7). The subunit structure of a transcription factor can be analysed by attempting to form heterodimers between the full-length factor and either truncated (20) or extended (21) forms of the factor that retain the capability of binding DNA. These experiments can be performed using an *in vitro* transcription/translation system (Promega) to synthesize the factor and these mutants. By mixing the translation products or cotranslating their mRNA, the subunit composition can be assayed by gel retardation (22). If a factor binds as a monomer, two retarded complexes will be formed, corre-sponding to the full-length factor and either a truncated or an extended form. However, if the factor forms a dimer, then three retarded complexes will be observed. There will be two homodimers and a heterodimer of intermediate mobility, which contains one molecule of full-length factor and a molecule either of the truncated or extended factor.

4.2 Analysis by chemical cross-linking

The monomer/dimer structure of a factor and interactions between transcrip-tion factors can be analysed directly by chemical cross-linking. If a factor exists as a dimer, cross-linking of the two subunits will double its molecular weight as observed under denaturing conditions such as SDS-PAGE. Cross-linking compounds react with chemical groups in proteins, including those found in the amino acid backbone and those found in the side chains. They form a covalent bond between two reactive groups and thus stably link them. Owing to the small size of these compounds, the cross-linked chemical groups must be in close proximity, inferring a direct interaction between their cognate polypeptides. Thus cross-linking is a sensitive method for directly assaying protein–protein interactions, such as dimer formation, and other interactions. There are many different chemical cross-linkers that are characterized and grouped according to their cross-linking properties (23). Each compound of this extensive array of cross-linkers has distinctive chemical characteristics and is used to cross-link different chemical groups separated by varying distances. They include homobifunctional and heterobifunctional compounds, some of which are cleavable or reversible.

Extensive review of these cross-linking reagents is beyond the scope of this chapter. The Pierce catalogue contains an extensive list of cross-linking reagents and protocols for their use. However, we will give an example of this powerful technique. The choice of a particular cross-linker depends on the groups that you want to cross-link. However, with uncharacterized transcription factors, the information to make the best choice is not available, thus more general cross-linkers are used. The best results are achieved by trial and error. For example, Aranyi *et al.* (24) used several related compounds, bisimidates (e.g. dimethyl succinimidate, dimethyl adipimidate, and dimethyl-pimelidate) to cross-link the progesterone receptor to show it exists as a dimer in solution. However, only dimethylpimelidate (DMP) successfully cross-linked the progesterone receptor (PR). DMP (Sigma) is a homobifunctional imidoester which forms a covalent bond between opposing primary amines.

Protocol 7. Cross-linking transcription factors with DMP

1. Make the following additions:
- 0.1 volumes of 2.2 M triethanolamine (pH 8)
- 0.2 volumes of 100 mM DMP/HCl (100 mM DMP/HCl in 200 mM ethanolamine (pH 8), adjusted with NaOH)
- 100–200 ng of transcription factor
- make up the volume to 30 μl with TESH buffer (10 mM Tris, 1 mM Tris, 1 mM EDTA, 12 mM 1-thioglycerol (pH 7.6)).

2. Incubate the reaction for 30 min at room temperature.

3. Stop the reaction with 0.1 volumes 200 mM ethanolamine.

4. Add an equal volume of 2 × SDS loading buffer and boil the sample for 2 min.

Protocol 7. *Continued*

5. Run the sample on a 7.5% polyacrylamide-SDS gel with a stacking gel.

6. Transfer the proteins to nitrocellulose or Zeta-Probe membrane.

7. Detect the cross-linked factors by Western immunoblot analysis (see *Protocol 3*).

4.3 Analysis of co-operative binding between dimers bound to adjacent response elements

Often enhancer elements act synergistically to enhance transcription. This phenomenon has been studied extensively using transient transfections and *in vitro* transcription. As an example, tandemly linked progesterone response elements (PREs) upstream of a basal heterologous promoter fusion gene (TK-CAT) confer synergistic progesterone inducibility when compared with induction from single PREs (25). This observed synergism could be due to co-operative binding of two PR dimers to adjacent elements. To determine if PR dimers bind co-operatively to tandem PREs, varying concentrations of PR were assayed by gel retardation and two retarded complexes were observed, II and IV (see *Figure 3*). Methylation interference showed they were formed by binding of a single dimer and two dimers to the probe, respectively. At low concentrations of receptor, only complex II was detected; however, at increasing concentrations complex IV was detected. With increasing receptor concentrations, the formation of complex IV was progressively enhanced to become the predominant protein–DNA complex. The levels of complexes II and IV and free DNA can be quantitated using a beta-scope 603 blot analyser (Betagen, Waltham, MA), phospho-imaging, or excision of the respective bands and scintillation counting. Quantitation showed that complex II reached approximately 10% of bound DNA, but no higher. In contrast, complex IV is present only at higher concentrations of receptor, where it rapidly becomes the predominant form. Analysis of the binding indicates that co-operative DNA binding displays sigmoidal kinetics (see *Figure 3*) (25). If no co-operativity is observed between the binding of the dimers, then complex II will reach 50% before the appearance of complex IV.

5. Identification and initial characterization of non-DNA-binding transcription factors

Interaction of a transcription factor with itself or other transcription factors to form DNA-binding dimers is relatively easy to detect and analyse. In contrast, the interaction of a transcription factor with a non-DNA-binding transcription factor is not as easy to detect. However, techniques have been

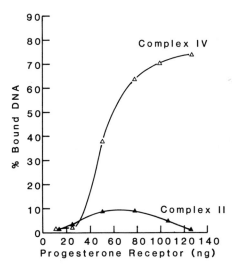

Figure 3. Quantitation of the two mobility shift complexes II and IV at different receptor concentrations and plotted the percentage bound in complexes II and IV.

developed to study these interactions. The requirement for a non-DNA-binding transcription factor is generally discovered during the purification of the DNA-binding transcription factor by a loss of transcriptional activity as the two factors are separated. Transcriptional activity can be restored by mixing two different fractions. In contrast to a requirement for an additional DNA-binding factor, the other fraction required for transcription activity will not bind DNA specifically, as determined by gel mobility shift assays and DNase I footprinting. The interactions between DNA-binding and non-DNA-binding transcription factors can be either strong or weak and necessitate different methods of analysis depending on their strength. Strong interactions are stable under the conditions of native electrophoresis and can be studied by gel mobility shift analysis and also by DNase I footprinting (see Chapters 1 and 2). Weak interactions are unstable and fall apart under the conditions employed for native gel electrophoresis and thus necessitate less stringent assays. These methods include immunoprecipitation, chemical cross-linking, kinetic analysis, or precipitation with glutathione-S-transferase (GST)–transcription factor fusion proteins.

The requirement for a non-DNA-binding factor is generally discovered fortuitously during the purification of a DNA-binding factor. There may be a significant or total loss of transcriptional activity as the DNA-binding transcription factor is progressively purified and assayed. An authentic loss of activity indicates the requirement for an additional transcription factor in addition to the DNA-binding factor. DNase I footprinting and gel retardation can be employed to determine if it is capable of binding specifically to DNA. The requirement of a DNA-binding transcription factor for a non-DNA-binding partner for transcriptional activity is not predictable. These requirements are peculiar and individual to each pair studied, thus it is difficult to describe protocols for the initial identification and analysis of non-DNA-binding transcription factors.

The first indications of a requirement for a non-DNA-binding transcription factor may not be as dramatic as a total loss of activity, it may first manifest itself as a non-alignment of DNA-binding and transcriptional activities in chromatographic fractions, as was observed for COUP-TF (a specific DNA-binding factor) and S300-II (a non-DNA-binding transcriptional factor) when chromatographed over a Sepharose S300 column (26). The two activities (DNA binding and stimulation of transcription) overlapped, but the peaks of each activity were not co-incident. Gel mobility shift assays and DNase I footprinting can be used to determine whether one or both peaks contain a specific DNA-binding activity. Where a requirement for a non-DNA-binding factor can be established, additional chromatographic steps can be used to further separate the two activities, leading to a total loss of transcriptional activity from the specific DNA-binding activity. COUP-TF and S300-II fractions containing the DNA-binding and non-DNA-binding activities, were rechromatographed on a smaller 33 ml Sephacryl S-300 column to obtain a better resolution and separation of the two activities (see *Figure 4*). The fractions collected from this column, containing the COUP-specific DNA-binding activity, were incapable of supporting transcriptional activity in the reconstituted system. However, when a fraction containing the non-DNA-binding factor, i.e. No. 57, was added to these assays, a peak of transcriptional activity was detected around fractions 41 and 42 (see *Figure 4*), which corresponded to the peak of DNA-binding activity. Fraction 57 was chosen to avoid COUP-TF contamination, as it was located on the side of the peak distal to

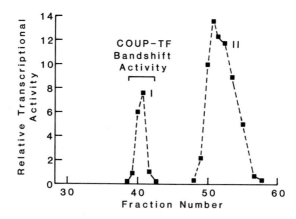

Figure 4. Dependence of the ovalbumin promoter transcription on two factors, COUP-TF I (peak I) and S300-II (peak II).

the DNA-binding activity. Similarly, the S300-II fractions supported only a low level of transcriptional activity; addition of fraction 41, containing COUP-specific DNA-binding activity most distal from the non-DNA-binding activity required for transcription, elicited a dramatic increase in transcriptional activity. Thus the second Sephacryl S-300 chromatographic step achieved separation of the specific binding and non-DNA-binding transcription factors and clearly demonstrated that transcriptional activity was dependent on both factors. Complementation of chromatographic fractions to the DNA-binding activity will lead to the restoration and identification of the fractions containing the non-DNA-binding activity. This outline highlights the approaches used to identify and study non-DNA-binding transcription factors. Such an initial characterization of the requirement for a non-DNA-binding factor establishes a reconstituted assay system for the non-DNA-binding factor, which will be required for subsequent purification and analysis of its function.

5.1 Assay of direct interactions between DNA-binding and non-DNA-binding transcription factors

Having established a functional requirement for a non-DNA-binding factor, for the transcriptional activity of a DNA-binding factor, the next step is to analyse this functional interaction. As it does not bind the response element, the non-DNA-binding transcription factor must interact directly, in some fashion, with the factor bound at this site. It may also interact with the general transcriptional machinery, by bridging or adapting the binding of the DNA binding transcription factor with a functional response from the RNA polymerase II transcriptional machinery. Thus the interactions of the DNA-binding and non-DNA-binding transcription factors needs to be studied. Interactions between factors can be analysed by such methods as immunoprecipitation, chemical cross-linking, or co-precipitation with GST fusion proteins of its putative partner, or using 'supershift' gel mobility shift assays, or kinetic analysis of transcription factor and DNA-binding stability. Generally, more than one technique is employed in the analysis of direct interactions, and the particular choices are guided by the strength of the observed interaction.

5.2 'Supershift' gel mobility shift analysis

If the interaction between the DNA-binding transcription factor and the non-DNA-binding transcription factor is sufficiently strong, it can be indirectly detected in a conventional gel mobility shift assay. The non-DNA-binding transcription factor, by its nature, will be incapable of binding to the probe, however, it will bind to the DNA-binding transcription factor bound to the probe, further retarding the migration of the complex during electrophoresis (see Chapter 1). This 'supershift' is similar in nature to that observed upon addition of specific antibodies. The conditions for the 'supershift' have to be determined empirically for each pair of DNA- and non-DNA-binding transcription factors, however, the interaction generally occurs under the transcription conditions used to assay their functional interaction (27).

Protocol 8. 'Supershift' gel mobility shift assay

1. Mix the partially purified DNA-binding and non-DNA-binding transcription factors with the labelled DNA fragment or oligonucleotide with non-specific competitor (0.5 μg *Hin*fI-digested pBR322 or 1 μg poly(dI-dC)·poly(dI-dC)) under transcription conditions previously determined.

2. Incubate at room temperature for 15–30 min.

3. Load the samples on to a native 4 or 5% polyacrylamide gel (the acrylamide percentage depends on the probe size).

4. Electrophorese for 1.5–3 h at 160–180 V in 0.5 × TBE.

5. Transfer the gel to Whatman 3MM paper and dry the gel at 80°C for 30 min.

6. Autoradiograph the dried gel with Kodak XAR-5 X-ray film with two Dupont Cronex Lightening Plus intensifying screens at −70°C.

5.3 Analysis of the dissociation of transcription factor– DNA complexes in the presence and absence of the non-DNA-binding transcription factor

If the interaction between the specific DNA-binding and the non-DNA-binding transcription factors is insufficiently strong, they will not be stable enough to detect as a 'supershift' in a gel retardation assay. However, non-DNA-binding factors affecting the DNA-binding stability can be assayed in a gel retardation assay by analysing the stabilization of the binding of the DNA-binding transcription factor to its response element. Stabilization of binding by the non-DNA binding transcription factor can be determined and quantified by analysing the 'off rate' of the DNA-binding transcription factor after it has bound to the response element. The 'off rate' is determined by challenging the transcription factor-labelled DNA complex with unlabelled response element in a kinetic experiment (28). The amount of unlabelled DNA added must be sufficient to eliminate completely the gel mobility shift if co-incubated with the probe and factor. Thus the amount used will vary from factor to factor and will depend on its DNA-binding affinity. The amounts of transcription factor and labelled response element used in each reaction in this experiment should be similar to those used to detect the transcription factor by gel mobility shift assay and should be incubated under similar conditions (see Chapter 1).

Protocol 9. Analysis of the dissociation of transcription factor–DNA complexes in the presence and absence of non-DNA-binding transcription factor

1. Mix the partially- or affinity-purified transcription factor with ^{32}P-labelled DNA fragment or oligonucleotide under conditions previously determined for a gel mobility shift assay (Chapter 1). Carry out one reaction for each time point. The reaction at each time point should be duplicated using an equivalent amount of BSA instead of transcription factor as a control for non-specific stabilization. (Non-specific stabilization is sometimes observed when a highly purified factor is mixed with other fractions or proteins.)

2. Incubate at room temperature for 10 min to allow the transcription factor to bind to the response element.

3. Add a 50- to 200-fold molar excess (depending on the DNA-binding affinity of the individual factor) of unlabelled specific competitor DNA fragment or oligonucleotide to each reaction after 10 min.

4. Terminate each reaction at specified time points by loading on to the gel and applying a very low voltage (this separates the bound and unbound probe immediately and stops the binding reaction). The addition of un-labelled competitor and the termination of the reaction should be accurately timed and spaced between reactions to produce an accurate and reproducible time curve. Typical kinetic time points may include 0, 2, 4, 8, 12, 16, 24, 32, and 60 min.

5. Analyse the reactions by native electrophoresis in a 4% or 5% polyacrylamide gel (depending on the probe size, see Chapter 1) and autoradiography as in a typical gel mobility shift assay.

6. Scan the autoradiograph with a densitometer, or scan the gel by phosphoimaging or with a beta scanner to quantitate the radioactivity in the retarded factor–DNA complexes.

7. Plot the amount of undissociated factor–DNA complex versus time to determine the $t_{1/2}$ of the dissociation of the complexes in the presence and absence of the non-DNA-binding transcription factor (see *Figure 2* for example).

5.4 Analysis of direct interactions by co-immunoprecipitation

Co-immunoprecipitation is a powerful technique that can detect direct interactions between two factors (29). It relies on the specificity of antibodies to recognize one of the components and protein A-Sepharose to precipitate the cross-linked immunocomplexes. Protein A-Sepharose is commercially available (Pharmacia). If the interaction between two factors is sufficiently strong, immunoprecipitation of one will bring down the other. However, if this interaction is not sufficiently strong, it can be stabilized by using a reversible chemical cross-linker such as dithio*bis*(succinimylpropionate) (DSP) (24, 30). It is a homobifunctional cross-linking reagent which reacts with primary amine groups. It can be cleaved with excess thiol and is thus reversible under denaturing SDS-PAGE conditions in the presence of β-mercaptoethanol. Thus SDS-PAGE separates the cross-linked species into their individual components. This procedure is especially useful if the factors have been cloned and can be labelled with [^{35}S]methionine by *in vitro* transcription/translation, or if antibodies are available. A drawback of co-immunoprecipitation is that non-specific entrapment of proteins can occur, especially when using polyclonal antibodies, which recognize many antigens leading to highly cross-linked aggregates. Controls are required to distinguish between specific co-immunoprecipitation and non-specific entrapment. Use another factor which is known not to react with the immunoprecipitation target. The immunoprecipitated pellets are washed several times with buffer to remove non-specific proteins, however, the stringency of these washes can be increased by adding small amounts of non-ionic detergents (e.g. 0.5% Nonidet P-40 (NP40) or 0.1% Triton X-100). In order to identify the co-precipitating factor, an antibody raised against it is required for Western immunoblot analysis, unless a clear identification can be made on the basis of molecular size, or the cloned protein can be labelled with [^{35}S]methionine during *in vitro* translation of its mRNA.

Protocol 10. Detection of direct interactions between factors by co-immunoprecipitation

1. Pre-incubate the partially purified DNA-binding and non-DNA-binding transcription factors in a 100 μl reaction volume under transcription conditions at a concentration of approximately 5 μg/ml for 1 h at 4°C.

2. Add 5 μg primary antibody (mouse monoclonal) or rabbit polyclonal antibody to the reaction. The final antibody concentration is approximately 50 μg/ml, but will depend on the individual affinity and titre of antibodies used.

3. Incubate for 2 h at 4°C with gentle agitation.

4. If a mouse monoclonal antibody has been used, add the secondary antibody, rabbit anti-mouse IgG, to a final concentration of 100–200 μg/ml. Incubate for 2 h at 4°C with gentle agitation. If using a rabbit polyclonal antibody, go directly to step 5.

5. Add 100 μl protein A-Sepharose 1:1 suspension in TE buffer (pH 7.5). Incubate for 1 h at 4°C with gentle agitation.

6. Microcentrifuge at 11 000 × *g* for 5 min at 4°C.

7. Carefully aspirate the supernatant and wash the pellet three times with 1 ml 20 mM Tris–HCl (pH 7.4), 100 mM NaCl, to remove non-specifically bound proteins.

8. Resuspend the pellet in 50 μl 1 × SDS-sample buffer, boil for 2 min, load on to a polyacrylamide-SDS gel (7.5% to 10%, depending on protein size) with a stacking gel and electrophorese (6).

9. The co-immunoprecipitation of the non-DNA-binding factor can be detected by Western transfer and immunoblot analysis using an antibody raised to the non-DNA-binding factor or by autoradiography if it has been labelled with [^{35}S]methionine.

Protocol 11. Stabilization of weak interactions with DSP

1. Incubate the transcription factors in transcription buffer[a] at 37°C for 30 min in a total volume of 50 μl, to allow association.

2. Add DSP to a final concentration of 2 mM.

3. Incubate at room temperature for 1 h.

4. Add ethanolamine to a final concentration of 0.1 M, to stop the cross-linking reaction.

5. Immunoprecipitate the cross-linked species (see *Protocol 10*).

6. Reduce and reverse the cross-linked immune complexes by boiling in 3 × SDS-PAGE buffer containing 15% β-mercaptoethanol.

7. Analyse by SDS-PAGE (6) and fluorography, if a [^{35}S]methionine-labelled factor has been used, or by Western analysis.

[a] Buffers used with DSP should be free of Tris as it contains primary amine groups which will quench the DSP reaction.

5.5 Analysis of direct interactions by co-immunoprecipitation with glutathione-*S*-transferase fusion expression and purification system

A recent innovation in the *E. coli* expression of heterologous proteins is the GST fusion expression and purification system (31). This system requires the in-frame insertion of cloned proteins into the 26 kDA GST protein. This allows isopropyl-β-D-thiogalactopyranoside (IPTG)-inducible high-level expression of a stable protein, from the *Lac* promoter, in *E. coli*. This GST-X fusion protein can be purified to near homogeneity using a one-step glutathione agarose batch or column purification procedure. It can be released from the glutathione agarose by addition of glutathione. In addition to the preparative uses of this technique, it also lends itself to analytical endeavours, such as the rapid assay of direct interactions between proteins (32). The purified GST–DNA-binding factor fusion protein can be mixed with fractions of partially purified non-DNA-binding factor. If an interaction occurs between these two factors, the non-DNA-binding factor can be co-precipitated with the GST-X fusion protein using glutathione agarose beads. A drawback of this procedure is that one of the interacting components has to be cloned and the other must be characterized sufficiently to positively identify its presence in the co-precipitate. This generally requires an antibody.

Protocol 12. Preparation of GST-X fusion extracts from *E. coli*

1. Subclone the cDNA of interest, in-frame into a GST expression vector (e.g. pGEX Pharmacia).

2. Inoculate a 10 ml LB broth (Ampicillin, 50 mg/ml) with a single recombinant colony. (Extracts have also to be made of the GST protein from the GST vector alone, as a control.)

3. Incubate at 37°C with shaking until the OD_{600} reaches 0.7 to 1 units.

4. Add IPTG (100 mM stock solution) to a final concentration of 0.5 mM to induce the fusion protein.

5. Incubate the culture for an additional 3 h at 37°C with shaking.

6. Centrifuge the culture at 2000 × g for 5 min at 4°C.

7. Resuspend the bacterial pellet in 1/50 of the original culture volume of lysis buffer (60 mM KCl, 20 mM Hepes–KOH (pH 7.9), 2 mM DTT, 1 mM EDTA, lysozyme 4 mg/ml (added fresh)).

8. Lyse the bacteria by three freeze/thaw cycles using liquid nitrogen and a 37°C water bath, being careful not to allow the lysates to warm much above 4°C.

9. Centrifuge the cell debris at 150 000 × g (45 000 r.p.m.) in a 60Ti rotor for 30 min at 4°C.

10. Aspirate the supernatant and add glycerol to a final concentration of 20% (v:v).

11. Store the extracts at −80°C or in liquid nitrogen.

Protocol 13. Co-precipitation using GST-X fusion proteins and glutathione-agarose

1. Add 1 ml *E. coli* lysate containing the GST-factor fusion protein to 25 ml of glutathione-agarose gel swollen 1:1 (v/v) in NENT + M buffer (100 mM

NaCl, 1 mM EDTA, 20 mM Tris–HCl (pH 8), 0.5% NP-40, 0.5% (w/v) non-fat dry milk). Set up a parallel reaction containing 1 ml of *E. coli* lysate containing GST alone, as a control.

2. Incubate at 4°C, with gentle agitation, for 1 h.

3. Centrifuge at 4000 r.p.m. for 5 sec at 4°C in a microcentrifuge. Aspirate the supernatant and wash the gel 3 times with 1 ml NENT + M buffer.

4. Combine the following components:

   • GST-X fusion–glutathione-agarose precipitate or GST control

   • partially purified non-DNA-binding transcription factor

   • NENT + M buffer to bring the volume up to 0.5 ml

5. Incubate for 1 h at 4°C on a rocker.

6. Centrifuge at 4000 r.p.m. for 5 sec in a microcentrifuge, aspirate the supernatant and wash the pellet 5 times with 1 ml NENT + M buffer.

7. Add 50 μl 1 × SDS-sample buffer to each sample, boil for 2 min and run on a polyacrylamide-SDS gel (7.5–10%) (6). 5 μl of supernatant of the GST control from step 3 can be run as a standard on the gel.

8. The co-precipitated proteins can be analysed by Western analysis and visualized by immunoblotting, or the target protein can be radiolabelled with [^{35}S]methionine using a coupled *in vitro* transcription translation system (Promega), if it has been cloned.

References

1. Sagami, I., Tsai, S. Y., Wang, H., Tsai, M.-J., and O'Malley, B. W. (1986). *Mol. Cell. Biol.*, **6**, 4259.
2. Andrews, P. (1970). In *Methods of Biochemical Analysis*, Vol. 18 (ed. D. Glick), pp. 1–53. Interscience, New York.
3. Laue, T. M. and Rhodes, D. G. (1990). In *Methods in Enzymology*, Vol. 182 (ed. M. P. Deutscher), pp. 566–87. Academic Press, London.
4. Fischer, L. (1980). *Gel-filtration Chromatography*. Elsevier, Amsterdam.
5. Stellwagen, E. (1990). In *Methods in Enzymology*, Vol. 182 (ed. M. P. Deutscher), pp. 317–28. Academic Press, London.
6. Laemmli, U. K. and Favre, M. (1973). *J. Mol. Biol.*, **80**, 575.
7. Martin, R. and Ames, B. (1961). *J. Biol. Chem.*, **236**, 1372.
8. Siegel, L. M. and Monty, K. J. (1966). *Biochim. Biophys. Acta*, **112**, 346.
9. Chodosh, L. A., Baldwin, A. S., Carthew, R. W., and Sharp, P. A. (1988). *Cell*, **53**, 11.
10. Merril, C. R. (1990). In *Methods In Enzymology*, Vol. 182 (ed. M. P. Deutscher), pp. 477–88. Academic Press, London.
11. Rodriguez, R., Weigel, N. L., O'Malley, B. W., and Schrader, W. T. (1990). *Mol. Endocrinol.*, **4**, 1782.

12. Slater, G. G. (1968). *Anal. Biochem.*, **24**, 215.
13. Timmons, T. M. and Dunbar, B. S. (1990). In *Methods in Enzymology*, Vol. 182 (ed. M. P. Deutscher), pp. 679–88. Academic Press, London.
14. Garfin, D. E. (1990). In *Methods in Enzymology*, Vol. 182 (ed. M. P. Deutscher), pp. 425–41. Academic Press, London.
15. Weber, K. and Osborn, M. (1969). *J. Biol. Chem.*, **244**, 4406.
16. Chodosh, L. A., Carthew, R. W. and Sharp, P. A. (1986). *Mol. Cell Biol.*, **6**, 4723.
17. Lin, S.-Y. and Riggs, A. D. (1974). *Proc. Natl. Acad. Sci. USA*, **71**, 947.
18. Hager, D. A. and Burgess, R. R. (1980). *Anal. Biochem.*, **109**, 76.
19. Bagchi, M. K., Tsai, S. Y., Tsai, M.-J., and O'Malley, B. W. (1987). *Mol. Cell. Biol.*, **7**, 4151.
20. Kumar, V. and Chambon, P. (1988). *Cell*, **55**, 145.
21. Tsai, S. Y., Carlstedt-Duke, J., Weigel, N. L., Dahlman, K., Gustafsson, J.-A., Tsai, M. J., and O'Malley, B. W. (1988). *Cell*, **55**, 361.
22. Ladias, J. A. A. and Karathanasis, S. K. (1991). *Science*, **251**, 561.
23. Pierce catalogue and references therein.
24. Aranyi, P., Radanyi, C., Renoir, M., Devin, J., and Baulieu, E.-E. (1988). *Biochemistry*, **27**, 1330.
25. Tsai, S. Y., Tsai, M.-J., and O'Malley, B. W. (1989). *Cell*, **57**, 443.
26. Sagami, I., Tsai, S. Y., Wang, H., Tsai, M.-J., and O'Malley, B. W. (1986). *Mol. Cell. Biol.*, **6**, 4259.
27. Buratowski, S., Hahn, S., Guarente, L., and Sharp, P. A. (1989). *Cell*, **56**, 549.
28. Tsai, S. Y., Sagami, I., Wang, H., Tsai, M.-J., and O'Malley, B. W. (1987). *Cell*, **50**, 701.
29. Lee, W. S., Kao, C. C., Bryant, G. O., Liu, X., and Berk, A. J. (1991). *Cell*, **67**, 365.
30. Yang-Yen, H.-F., Chambard, J.-C., Sun, Y.-L., Smeal, T., Schmidt, T. J., Drouin, J., and Karin, M. (1990). *Cell*, **62**, 1205.
31. Smith, D. B. and Johnson, D. S. (1988). *Gene*, **67**, 31.
32. Kaelin, Jr., W. G., Pallas, D. C. DeCaprio, J. A., Kaye, F. J., and Livingston, D. M. (1991). *Cell*, **64**, 521.

Appendix

In vitro transcription assays

In vitro transcription assays are often used to assay fractions for activity during the preparation of a transcription factor. We therefore give here protocols for the preparation of transcriptionally active nuclear extracts and for the assay itself.

Protocol 14. Preparation of HeLa cell nuclear extracts

1. Harvest HeLa cells from culture by centrifuging at $1000 \times g$ at 4°C for 10 min in a Beckman JA20 or JA14 rotor, depending on culture volume.

2. Resuspend the cell pellet in 5–10 packed cell volume (PCV) of ice-cold phosphate-buffered saline (150 mM NaCl, 16 mM Na_2HPO_4, 4 M NaH_2PO_4 (pH 7.3)).

3. Centrifuge at $500 \times g$ at 4°C for 10 min.

4. Resuspend the cell pellet in $5 \times$ PCV of buffer A (10 mM Hepes (pH 7.9), 1.5 mM $MgCl_2$, 10 mM KCl and 2 mM DTT[a]).

5. Centrifuge at $500 \times g$ at 4°C for 10 min.

6. Aspirate the supernatant and resuspend the cells in $2 \times$ PCV of buffer A.

7. Transfer the cells to a glass Dounce homogenizer (B-type pestle). Lyse the cells with 10 strokes.

8. Examine the homogenate microscopically for maximum cell lysis (90% of cells should be lysed).

9. Centrifuge at $500 \times g$ at 4°C for 10 min. Carefully aspirate the supernatant and recentrifuge the pellet at $27\,000 \times g$ in a Beckman JA20 rotor at 4°C for 20 min to remove any residual supernatant.

10. Resuspend the nuclei pellet in $0.5 \times$ PCV of buffer C (containing 0.6 M NaCl) or $0.66 \times$ PCV of buffer C (containing 0.55 M NaCl) (buffer C: 20 mM Hepes (pH 7.9), 20% glycerol, 1.5 mM $MgCl_2$, 0.2 mM EDTA, 2 mM DTT[a]).

11. Transfer to an appropriately sized glass Dounce homogenizer (B-type pestle) and lyse the nuclei with 10 strokes.

12. Transfer the lysate to 50 ml polypropylene centrifuge tubes and cover with para film. Incubate on ice for 30 min, vortexing gently at 5 min intervals.

13. Centrifuge at $27\,000 \times g$ for 30 min at 4°C in a JA20 rotor.

14. Recover the supernatant (nuclear extract) and place in a dialysis bag. (M_r cut-off 12 000–14 000 Da.) Dialyse against 100 volumes of buffer D (20 mM Hepes (pH 7.9), 20% glycerol, 0.1 M KCl, 0.2 mM EDTA, 5 mM $MgCl_2$, 2 mM DTT[a]) for 4 h, with one change of buffer.

Protocol 14. *Continued*

15. Centrifuge the nuclear extract at $27\,000 \times g$ for 20 min at 4°C in a JA20 rotor. Recover the supernatant, aliquot and freeze at −80°C.

ᵃ DTT is added fresh.

Protocol 15. *In vitro* transcription using the G-free cassette

1. Make the following additions to an Eppendorf tube in a final reaction volume of 25–30 µl:

- 12.5–200 ng of control and test template DNAs*ᵃ*
- 60 mM KCl
- 12 mM Hepes (pH 7.9)
- 5 mM MgCl$_2$
- 12% glycerol
- 0.12 mM EDTA
- 2 mM DTT
- 500 µM ATP, CTP
- 20 µM UTP
- 2 µCi/reaction [^{32}P]UTP (800 Ci/mM)
- 10 IU/reaction Ti nuclease
- 1 µmol/reaction 3′-*O*-methyl GTP
- herring sperm DNA (0.5–1 µg/reaction)
- 3–5 µl nuclear extract

Mix by gentle vortexing and spin down briefly. Start the reaction by adding the ribonucleotide mix and incubate at 30°C for 45 min.

2. Terminate transcription by adding 85 µl of 30 mM Tris–HCl (pH 8.0), 10 mM EDTA, 0.5% SDS, 120 µg/µl yeast tRNA, and 400 µg/ml proteinase K.

3. Incubate at 37°C for 30 min.

4. Add 200 µl urea stop solution per reaction (10 mM Tris–HCl (pH 8.0), 1 mM EDTA, 8 M urea).

5. Extract twice with an equal volume of phenol (buffered to pH 5.0 with 10 mM sodium acetate): chloroform:isoamyl alcohol (25:24:1, by vol.).

6. Aspirate the upper aqueous phase and add one-tenth volume 3 M sodium acetate (pH 5.2), 10 µg yeast tRNA and 2.5 volumes of absolute ethanol. Incubate in a dry ice/ethanol bath for 30–40 min.

7. Centrifuge at 13 000 × *g* for 60 min at 4°C. Wash the pellet once with 80% ethanol (−20°C).

8. Centrifuge for 5 min, aspirate the supernatant and dry the pellet.

9. Resuspend the pellet in 4 μl of 80% deionized formamide, 1 mM EDTA. 1 ng/ml xylene cyanol, 1 ng/ml bromophenol blue. Heat at 95°C for 5 min and cool quickly on ice.

10. Run the samples on a 7% acrylamide urea sequencing gel. Electrophorese at 1800 V.

11. Fix the gel in 10% methanol and 10% acetic acid for 5 min and dry the gel.

12. Autoradiograph overnight at −70°C with two intensifying screens.

---

[a] The amount of DNA added depends on the strength of the promoter.

4

Purification and cloning of transcription factors

R. H. NICOLAS and G. H. GOODWIN

1. Introduction

If you have just identified a protein binding to a novel regulatory sequence, the race is on to clone the cDNA encoding the protein. Most practitioners in this field would first attempt direct screening of expression libraries with the DNA-binding site (see Chapter 5) before embarking on the protein purification/peptide sequence/oligonucleotide screening approach. However, in many cases, direct screening of libraries has not proved successful or has not been possible (for example when the protein is part of a multiprotein complex and more than one protein is required for DNA binding) and protein purification may then have been the only method available. The purification of DNA-binding proteins has been remarkably successful with the development of sequence-specific DNA affinity chromatography and the band-shift technique for rapidly analysing chromatography fractions. With the recent advances in protein-sequencing techniques (e.g. sequencing peptides immobilized on blots following electrophoretic separation), the protein purification approach has become less arduous and one can expect to get good sequence from 50 pmol of protein.

One of the major difficulties faced by the protein purifier is the abundance of the transcription factor in the cell. Eukaryotic transcription factors are present in the cell nucleus in widely varying quantities, ranging from 10^3 molecules per cell for some retinoic acid receptors to well over 10^5 per cell for the GATA-1 factor in erythroid cells. If at least 100 pmol of pure protein are required for production of peptides for sequencing, it is apparent that for a protein of average abundance at 10^4 molecules per cell, more than 10^{11} cells of starting material are required, assuming 5% recovery of protein.

The successful isolation of protein also required that the molecular weight of the protein has been determined (for monitoring the final stages of purification of the protein) and that a DNA sequence with high affinity and specificity for the protein has been identified. As a rough guide, the ratio of the strengths of binding to specific versus non-specific DNA sequences should be at least 100. This can be determined using competitive DNA-binding analyses in the band-shift assay. The molecular weight of the protein can be determined by two-dimensional electrophoretic techniques (1), by photoaffinity labelling (2), or by Southwestern blotting techniques (3).

   In this chapter, we describe the techniques used to purify moderately abundant factors, as illustrated by the purification of a zinc-finger factor termed CTCF, a ubiquitous protein which binds to a GC-rich element in the promoter region of the chicken *c-myc* gene (4), and chicken GATA1 (formerly termed EF1, NF, E1, EryF1) a protein expressed in erythroid cells, mega-karyocytes, and platelets (1, 5). This is outlined schematically in *Figure 1*. In these purifications, DNA-cellulose chromatography is carried out to isolate DNA-binding proteins prior to DNA affinity chromatography. Other chro-matographies commonly used to achieve partial purification before affinity chromatography include heparin-sulphate, phosphocellulose or ion-exchange. In our view, gel filtration rarely has sufficient resolution or capacity to make it worthwhile. Following an affinity chromatography step, final purification is achieved by further cycles of affinity chromatography or by reverse-phase chromatography.

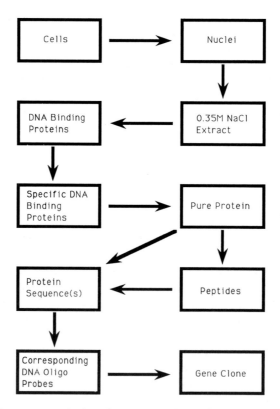

Figure 1. The steps required to clone a transcription factor via protein isolation.

2. Buffers and solutions

When the aim of purifying a transcription factor is to obtain some primary sequence of the protein, it is important that all solutions should be scrupulously clean. Any glassware used should be acid-washed by soaking overnight in nitric acid (one part nitric acid, two parts HPLC-grade water), then rinsed four times in HPLC-grade water and dried in an oven at 150°C. Sterile disposable plasticware is a useful alternative, and tissue-culture flasks make good reagent bottles. Only high-quality chemical reagents should be used, and stock solution should be filtered through 0.22 μm filters. Many of the buffers contain glycerol for protein stability, but it is very difficult to measure viscous reagents by volume, so glycerol is added by weight. It is also a good idea to make a glycerol-containing stock solution such as Ns (*Table 1*) which can be used to make all the buffers for DNA chromatography.

3. Preparation of nuclear extract

Consideration should be given first to which tissue or cell line will give the greatest yield. In general, tissue culture is expensive, but relatively pure nuclei can be prepared in high yield in a couple of steps using just low-speed centrifugations. To prepare nuclei and obtain undegraded nuclear proteins from tissues, the starting material must be fresh. This may present difficulties when working with human tissue. Animal tissue may be more readily available, but it should be remembered that tissues do not contain a homogeneous cell population and only a few cell types may be expressing the protein of interest. Some tissues, such as thymus and spleen, may be gently disrupted in phosphate-buffered saline (PBS) and filtered through gauze to yield a crude cell suspension and then *Protocol 1* may be followed to prepare a nuclear extract. With tougher tissues, it is better not to prepare cells but prepare pure nuclei directly by centrifugation through high-sucrose gradient as described in *Protocol 2.*

The preparation of nuclei gives between 10- and 100-fold enrichment of a nuclear-localized protein, depending on the size of the cells. The cytoplasmic supernatant fractions should be assayed for activity because in some cases weakly bound factors may be lost from the nuclei. Some transcription factors may be susceptible to denaturation by the detergent treatment used to lyse the cell membrane, in which case cells may alternatively be lysed by hypotonic shock (6). In many cases, lower concentrations of Triton are sufficient to lyse cells and this can be investigated for each cell type, but for optimum yields it should be ensured by microscopic examination that all the cells are lysed.

Transcription factors are then extracted from the nuclei with a buffer containing 0.3 M NaCl. The highest concentration of salt that can be used without extracting histone H1 is 0.35 M NaCl. Extraction of H1 at higher salt concentrations results in chromatin decondensation, nuclear lysis, and

Table 1. Solutions

(a) *Stock solutions*

    200 mM Hepes–NaOH (pH 7.9)
    200 mM EDTA (pH 8.0)
    1 M Tris–HCl (pH 7.6)
    1 M MgCl$_2$
    4 M NaCl
    2 M (NH$_4$)$_2$SO$_4$
    10% Triton X-100 (Pierce)
    10% Brij-35 (Pierce)
    Glycerol
    1 M DTT   Store in aliquots at −20°C and add to solutions
                                     on the day of use

(b) *Stock of protease and phosphatase inhibitors*

    50 mM PMSF (phenylmethylsulphonyl fluoride). Dissolve in propan-2-ol and store
      in a dark bottle at 4°C for up to 3 months, as long as it is kept free of water
    50 mm benzamidine in water
    100 μg/ml leupeptin in water
    100 μg/ml aprotinin in water
    100 μg/ml pepstatin in methanol
    200 mM levamisole in water
    1 M β-glycerophosphate in water

All these solutions are 100 × concentration and should be added to the pre-cooled
solutions on the day of use. All except PMSF can be stored as aliquots at −20°C for
up to 3 months or at 4°C for 1 week.

(c) *Solutions for preparation of nuclear extracts*

TM	10 mM Tris–HCl (pH 7.6), 5 mM MgCl$_2$
TTM	10 mM Tris–HCl (pH 7.6), 5 mM MgCl$_2$, 0.5% Triton X-100
PBS	Dulbecco's phosphate-buffered saline. Dissolve 1 tablet (OXOID) in 100 ml or dissolve 0.2 g KCl, 0.2 g KH$_2$PO$_4$; 8.0 g NaCl and 2.16 g Na$_2$PO$_4$.7H$_2$O in 1 litre. The pH should be 7.3–7.4
N100	See *Table 1d*
NDil	See *Table 1d*

All the above solutions should have the inhibitors added just before use.

(d) *Buffers required for DNA chromatography*

Buffer	Reagents required in millilitres to make 100 ml of buffer					
	Ns[a]	Water	4 M NaCl	2 M (NH$_4$)$_2$SO$_4$	1 M DTT	Inhibitors
N100	50	40.4	2.5	–	0.1	7 × 1
N200	50	37.9	5.0	–	0.1	7 × 1
N1000	50	17.9	25	–	0.1	7 × 1
NDil	75	17.9	–	–	0.1	7 × 1
E250	50	30.4	–	12.5	0.1	7 × 1
N100/50[b]	50	10.4	2.5	–	0.1	7 × 1

[a] Ns is 40% v/v glycerol (= 50% w/v glycerol), 10 mM MgCl$_2$, 0.2 mM EDTA, 40 mM Hepes (pH 7.9), 0.1% Brij-35.
[b] N100/50, make up to volume with glycerol (37.5 g).

formation of a gel that can only be effectively removed by ultracentrifugation. H1 also causes problems during the subsequent chromatography steps since it is very abundant and will bind to the DNA-cellulose and bind non-specifically to the DNA affinity columns. Although H1 can be removed by ion-exchange chromatography, it is better to avoid extracting the H1 in the initial nuclear extraction, provided the factor of interest is extracted in high yield by 0.3–0.35 M NaCl. Since the nuclear extraction buffer contains other components which contribute to the ionic strength, the NaCl should be added to only 0.3 M final concentration to minimize H1 solubilization.

Following extraction of the nuclei, the NaCl concentration of the extract is reduced either by dilution or dialysis before loading on to the DNA-cellulose column. This causes precipitation of proteins, and some transcription factors may come out of solution at this point. Since the solubility is usually inversely proportional to NaCl concentration, the highest NaCl concentration that will still allow binding to the DNA-cellulose column should be determined. Conversely, if the factor remains soluble at low salt concentration, then by optimizing this precipitation a two- to threefold purification can be achieved. The use of a non-ionic detergent such as Brij-35 does keep some factors soluble, but it may also be detrimental to other binding activities.

Ammonium sulphate fractionation can be included before the DNA-cellulose chromatography. This has the advantage of concentrating the protein and removing some nucleic acid, but since the precipitate has to be redissolved in buffer and then dialysed against N100 buffer before loading on to column, there is little saving in the time taken for the preparation. Ammonium sulphate precipitation can be detrimental; e.g. in the case of GATA1, some activity was lost due to denaturation or incomplete dissolution of the protein precipitate. Interestingly, when 3.5 g $(NH_4)_2SO_4$ is added per 7.5 ml of 0.3 M nuclear extract, the majority of the proteins precipitate, leaving only the high mobility group (HMG) proteins in solution. Thus, after pelleting the ammonium sulphate precipitate by centrifugation ($40\,000 \times g$, 30 min), the HMG proteins can simply be prepared in an undenatured form by extensively dialysing the supernatant.

Protocol 1. Preparation of nuclear extract

1. Collect cells by centrifugation at $2000 \times g$.
2. Resuspend the cells in PBS, centrifuge $2000 \times g$ for 10 min.
3. Resuspend in 10 volumes of TTM buffer,[a] homogenizing with a loose-fitting Dounce glass homogenizer, five strokes is usually enough.
4. Check that cells have lysed by examining a methylene blue-stained sample under the microscope.
5. Centrifuge at $5000 \times g$ for 20 min.

Protocol 1. *Continued*

6. Resuspend the crude nuclei in 10 volumes TM buffer[a] and centrifuge at $5000 \times g$ for 10 min.

7. Resuspend in TM buffer. Use 1 ml for every 5×10^8 starting cells. Make the suspension 0.3 M NaCl by adding 4 M NaCl. Ensure the extract is mixed thoroughly whilst adding the 4 M NaCl. Leave at 4°C for 30 min and then centrifuge at $40\,000 \times g$ for 30 min.

8. (a) Dialyse the supernatant versus N100[a] buffer overnight, *or*
   (b) Add 2 volumes of zero salt dilution buffer (NDil).[a] Mix by gentle stirring. Leave on ice for 1 h.

9. Clarify the nuclear extract by centrifugation for 1 h at $40\,000 \times g$ and collect the supernatant. Freeze rapidly on solid CO_2 in aliquots or load directly on to the DNA-cellulose column.

[a] See *Table 1*.

Protocol 2. Preparation of nuclei from whole tissue, e.g. liver

1. Dice 100 g liver and homogenize for 2 min at low speed in a domestic blender or equivalent with 1 litre of TM buffer[a] plus 0.25 M sucrose.

2. Filter through four layers of surgical gauze.

3. Centrifuge at $2500 \times g$ for 15 min.

4. Discard the supernatant, add 1 litre of TTM buffer[a] containing 0.25 M sucrose and stir or very gently homogenize. Stir for 15 min and centrifuge as above.

5. Discard the supernatant. Gently homogenize the pellet with 5 volumes of 2.4 M sucrose in TM buffer and then centrifuge at $35\,000 \times g$ for 1 h.

6. Remove supernatant and resuspend the nuclear pellet in 70 ml of TM buffer.

[a] See *Table 1*.

---

4. DNA-cellulose chromatography

The calf thymus DNA-cellulose chromatography is a soft-gel technique and does not require any special apparatus apart from a peristaltic pump, column, fraction collector, and UV monitor. The sample should be loaded in a reasonably low ionic strength buffer; most DNA-binding proteins bind at 100 mM NaCl. In some cases (e.g. for the GATA1 factor), yields from the DNA-cellulose column can be low unless the detergent Brij-35 is included in the buffer. After loading, the elution from the column should be monitored by measuring the absorbance at 280 nm and washed until the absorbance returns to baseline. It is advisable not to stop the buffer flow for long during the chromatography as this results in an increase in absorbance, presumably due to some DNA eluting off the column. The DNA-binding proteins are eluted from the column at high ionic strength (E250 buffer, 250 mM $(NH_4)_2SO_4$), and usually all the protein is eluted in less than a column volume. There is usually little advantage in using more complex chromatographic elutions (i.e. using gradient or multiple salt steps). The ratio of bound to non-bound protein is usually 1:10 and, in the case of GATA1, an approximate yield of 70% is usually obtained.

Protocol 3. DNA-cellulose chromatography

1. Weigh out 10 g of DNA-cellulose (Sigma) and allow it to swell in 300 mM NaCl, 10 mM Tris–HCl (pH 7.6), 1 mM EDTA. Suspend in 100 ml of buffer and allow to stand for a few minutes. Remove the fines by aspiration. Pack the cellulose into a 20 ml column (1.6 mm diameter) and place in the 4°C cold room.

2. Wash column with 1 volume of 2 M NaCl, 10 mM Tris–HCl (pH 7.6), 1 mM EDTA at a flow rate of 0.2–0.5 ml/min.

3. Wash column with 1 volume 100 mM NaCl, 10 mM Tris–HCl (pH 7.6), 1 mM EDTA.

4. Wash column with at least 3 volumes of N100 buffer.[a]

5. Pump clarified nuclear extract on to the column. Large volumes can be loaded at a slow flow rate overnight. Approximately 200–400 mg of protein is obtained from 10^{11} cells.

6. Once loaded, wash the column with at least 2 column-volumes N100 buffer until the absorbance at 280 nm returns to background level.

7. Elute the DNA-binding proteins with E250 buffer.[a, b] Most of the protein is eluted in 1 column-volume as judged by absorbance at 280 nm.

8. Pool the fractions and dialyse against at least 20 volumes N200 buffer for a minimum of 5 h.

9. Flash-freeze in aliquots at −70°C to store or continue with affinity chromatography.

[a] See *Table 1*.
[b] Aliquots of fractions should be kept for band-shift analysis. Also, an estimate of the yield of protein is useful at this stage. This can be determined by the Bio-Rad protein assay with minimal interference by non-ionic detergents. Check that there is no activity remaining on the column following elution with E250 by eluting with a higher $(NH_4)_2SO_4$ concentration (e.g. 500 mM).

---

5. DNA affinity chromatography

The method has not varied significantly since originally developed by Kadonaga and Tjian (7) in 1986 when they first described the isolation of the transcription factor SP1. In this method, a large number of DNA-binding sites were linked to the column by first polymerizing an oligonucleotide containing a strong SP1-binding site and this was then covalently attached to the column matrix (Sepharose). One should aim at more than 2 nmol per millilitre of matrix. This maximizes the concentration of protein-binding sites in the column so that there are a large number of specific sites to compete with the competitor DNA that is required for selectivity. Commercially available cyanogen bromide (CNBr)-activated Sepharose 4B (Pharmacia) works well, provided it is not close to its expiry date (*Protocol 4*). Double-stranded DNA is very unreactive towards CNBr Sepharose 4B, so the oligonucleotide has to have an overhang of at least four base pairs at the end of the DNA and should contain a G base. ACGT is commonly the overhang used. In order to anneal the complementary strands of the oligonucleotide, the extinction coefficients of both strands must be calculated and their concentrations determined accurately. Great care should be taken that the two strands are annealed in equimolar amounts so that no single-stranded DNA remains to be attached to the Sepharose.

The binding site for the transcription factor should have been thoroughly characterized by band-shift analysis (see Chapter 1), footprinting, and related techniques (see Chapter 2). The sequence chosen for affinity chromatography should, if possible, bind only one factor specifically and should be the strongest site that has been found. It may be necessary to screen a number of potential sites and also generate mutants to find a binding site with high specificity and affinity. Ideally, the affinity column should bind the protein strongly enough so that it does not elute off the column below 0.5 M NaCl; if it elutes below 0.3 M NaCl, the protein will be very difficult to purify without a number of other steps in the purification procedure.

The selectivity of the method is determined by the immobilized polynucleotide and by the concentration and character of the competitor DNA. All the proteins loaded on to the column have the potential to bind to the column. The trick is to add as much competitor DNA as possible so that most non-specific DNA-binding proteins partition on to the mobile competitor DNA. In practice, one is limited in how much non-specific DNA can be added without precipitating some of the factor of interest. As a rule of thumb, one can add up to an equal weight of poly(dI-dC) as total weight of protein in the sample to be loaded on to the column. More complex DNA such as *E. coli* DNA is a more effective competitor but usually starts to compete for the specific protein at lower concentrations. Other synthetic DNA polymers that may help the factor bind more specifically can also be included (8). The sequence used to purify the CTCF factor also binds a polyG-binding protein and SP1 (4). Since band-shift analysis showed that poly(dG).poly(dC) would compete for the polyG-binding factor and not CTCF, this was included in the final mix and this prevented binding to the column. Typically, in the purification of CTCF, 16 ml of the E250-eluted fraction from the DNA-cellulose column (1 mg/ml protein) was mixed with 8 ml of poly(dI-dC) and 160 μl of poly(dG)·poly(dC) both at 1 mg/ml in 100 mM NaCl.

Non-specific binding of proteins to the affinity column can be due to several reasons. (a) Another protein(s) may bind to a site adjacent to or overlapping the sequence of interest. This problem can be corrected by band-shift analysis and redesigning the oligonucleotide bound to the column; remember that on ligation a new site may be formed. (b) Some proteins may have an affinity for Sepharose 4B. For example, the GATA1 factor binds to Sepharose and elutes gradually at salt concentrations between 100 and 300 mM NaCl. This may be solved by including a pre-column of Sepharose 4B before the affinity column. (c) Thirdly, there are proteins that bind to the DNA of the affinity column because they do not have a high affinity for the competitor DNA. The nature of these contaminants will depend on the nuclear extract, the DNA of the affinity column, and the competitor DNA. One such example are the single-strand nucleic acid binding proteins associated with nuclear hnRNA. During the purification of GATA1, four or five of these proteins consistently co-purified with GATA1. GATA1 could be separated from their contaminants by reverse-phase chromatography (see below), or by recycling through the affinity column, although this resulted in lower yields since contaminated side fractions had to be discarded. If there is evidence that these 40 kDa proteins are binding to the column, then they can be removed by passing the extract through a single-strand DNA-cellulose column after the double-stranded DNA-cellulose column (5). (d) It is possible that a DNA-binding protein has one or more non-DNA-binding proteins associated with it. Such large molecular complexes may be manifested by slowly migrating bands in band-shift analysis. Indeed, a slow broad band might hide a number of complexes. Cross-linking experiments with deoxybromouridine (2) can be used to determine the molecular weight and complexity of the DNA-binding components but may give no indication of the size and complexity of the non-DNA-binding components (see Chapter 3). It has recently been shown that a modification of a method relating mobility to molecular weight described by Ferguson (9) for proteins can be applied to the DNA–protein complexes (10). This type of approach may help in determining the overall composition of the complex. If one is sure that all the proteins bound to the column are components of the complex, then in principle they could all be isolated, sequenced, and hence cloned. In practice, one should consider raising antibodies to the complex to generate reagents that will aid in the purification and cloning, and help elucidate the structure of the complex.

The pre-column step is included in *Protocol 5* but is not mandatory; it may help to remove some of the contaminants from the final products. The

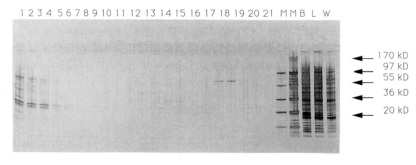

(A)

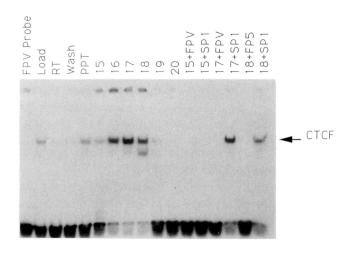

(B)

Figure 2. (A) SDS-PAGE analysis of fractions obtained following affinity chromatography of CTCF. 10 µl of each sample plus an equal volume of SDS loading buffer was electrophoresed on a Pharmacia Excel gel and stained with silver. Numbering of fractions starts with the start of the gradient. Most of the CTCF eluted at 500 mM NaCl in fractions 17 and 18. M, protein markers (BCL) loaded at 1 µg and 100 ng per band; B, the pooled protein from the previous DNA-cellulose chromatography; L, is the protein loaded on to the affinity column after adding competitor DNA and centrifuging; W, the wash fractions eluted with N200 buffer. (B) Band-shift analysis of CTCF DNA-binding activity in chromatography fractions. The ^{32}P-labelled DNA probe FPV (2 fmol) was incubated with various competitors and protein fractions and electrophoresed on a 4% polyacrylamide gel (1). From the left to right: probe alone; Load, protein loaded on to affinity column; RT, protein not bound to the column; Wash, protein eluted with N200 buffer; PPT, precipitation formed on addition of competitor DNA prior to chromatography (~5% of total CTCF); 15–20, fractions of the affinity chromatography. Fractions 15, 17, and 18 were also incubated with a 100-fold molar excess of unlabelled FPV probe or with the same amount of an oligonucleotide containing an SP1 binding site, as indicated.

pre-column matrix is usually Sepharose 4B with an immobilized polymerized oligonucleotide having a sequence unrelated to that bound by the factor being purified. In practice, we have used the run-through peak from one DNA-affinity-chromatograph column to load on to a second different DNA affinity column to purify a second transcription factor. Thus a number of factors were prepared from the same nuclear extracts. The run-through peaks are especially valuable as they have competitor DNA added and should be stored by fast-freezing and kept at −70°C until required.

Figure 2 shows an SDS-polyacrylamide gel and the band-shift analysis of the fractions from DNA affinity chromatography during the purification of CTCF from chicken tissue-culture cells. The contaminating hnRNA-binding proteins can be seen in the early fractions and elute from the column at approximately 300 mM NaCl. The sample loaded on the column had already been passed through a GATA1-binding column which effectively removed most of these contaminating proteins. CTCF eluted at 500 mM in a total volume of 4 ml. It can be seen from the analysis of *Figure 2* that a major protein of 130 kDa co-elutes with the DNA-binding activity. Since photo-affinity labelling of CTCF had characterized CTCF as a 130 kDa protein, it was assumed that the protein band seen on the SDS gel was the CTCF protein. Band-shift and footprinting analyses demonstrated that the purified protein had the same characteristics as the original CTCF in crude nuclear extracts. The protein was finally purified by reverse-phase chromatography as described in the next section.

Protocol 4. Preparation of DNA affinity columns

You will need:

- 10 × K buffer: 400 mM Tris–HCl (pH 8.0), 100 mM MgCl$_2$, 0.1 mM EDTA
- 10 × ligation buffer: 500 mM Tris–HCl (pH 7.8), 100 mM MgCl$_2$
- 5 M ammonium acetate
- 10 mM ATP (pH 7.5)
- 10 Ci/ml [^{32}P]ATP (Amersham)
- 100 mM potassium phosphate (pH 8.0)
- 1 mM HCl
- 1 M ethanolamine–HCl (pH 8.0)
- 1 M KCl, 1 mM EDTA (pH 7.6)
- 300 STE: 300 mM NaCl, 10 mM Tris–HCl (pH 7.6), 1 mM EDTA
- NAP-5 column (Pharmacia)
- CNBr-activated Sepharose 4B (Pharmacia)
- 0.22 μm 115 ml filter (Nalgene)

Protocol 4. *Continued*

1. Phosphorylate the two DNA oligonucleotides separately with poly-nucleotide kinase and ATP. For each strand mix the following:
   - 10 × K buffer $\qquad$ 10 µl
   - DNA oligonucleotide $\qquad$ 20 nmol
   - 10 mM ATP $\qquad$ 10 µl
   - 10 Ci/ml [γ-^{32}P]ATP $\qquad$ 0.1 µl
   - H$_2$O, to make up to 100 µl

   Incubate for 2 h at 37°C with 100 IU of T4 polynucleotide kinase.

2. Add 50 µl of 5 M ammonium acetate, followed by 300 µl ethanol. Leave on dry ice for 5 min and collect the precipitate by centrifugation. Wash the precipitate with 80% ethanol and dry under vacuum.

3. Dissolve the DNA oligonucleotides in 40 µl of water and mix them together. Add 10 µl of 10 × ligase buffer.

4. Heat to 90°C for 5 min and allow to cool to room temperature. Anneal at room temperature for at least 30 min.

5. Add and mix the following:
   - 10 mM ATP $\qquad$ 10 µl
   - 1 M dithiothreitol (DTT) $\qquad$ 1 µl
   - T4 DNA ligase $\qquad$ 8 IU

   and incubate for 16 h at 16°C in a water-bath in the cold room.

6. Extract the DNA with phenol/chloroform followed by chloroform.

7. Purify the DNA from any buffer components containing amino groups and ATP by gel filtration chromatography by either of these methods:

   (a) The aqueous phase can be loaded on to a Pharmacia NAP-5 gel-filtration column that has been equilibrated with 100 mM potassium phosphate buffer (pH 8.0). Collect 200 µl fractions and monitor the absorbance at 260 nm or use a beta counter to monitor the ^{32}P-labelled DNA.

   (b) Alternatively, a spinning column may be used (11). The matrix from a NAP-5 column may be packed into a 1 ml syringe and equilibrated with 100 mM potassium phosphate (pH 8.0).

8. Estimate the yield of DNA from its absorbance at 260 nm and retain an aliquot for scintillation-counting the ^{32}P. A sample should also be analysed by agarose gel electrophoresis to measure the extent of ligation which is usually between 10–30 oligonucleotides per chain.

9. Swell 1 g CNBr-activated Sepharose in 100 ml 1 mM HCl for 1 h.

10. Filter off the supernatant under a slight vacuum but do not allow the Sepharose to become dry.

11. Wash the Sepharose with a further 200 ml of 1 mM HCl.

12. Wash the Sepharose with 100 ml of water and transfer to a 15 ml screw-cap plastic tube. Allow the 3.5 ml of Sepharose to settle out and remove all but 1.0 ml of the supernatant.

13. Add 0.5 ml 100 mM potassium phosphate buffer (pH 8.0), immediately followed by the oligonucleotide solution from step 8. Rotate the tube on a roller for 20 h at room temperature.

14. Remove supernatant and add 100 ml of 1 M ethanolamine–HCl (pH 8.0) and rotate for a further 3 h.

15. Transfer to a filter and wash slowly (approximately 10 ml/min flow) with:
    (a) 100 ml 1 M ethanolamine–HCl (pH 8.0)
    (b) 100 ml 10 mM potassium phosphate buffer (pH 8.0)
    (c) 100 ml 1 M KCl, 1 mM EDTA (pH 7.6)
    (d) 100 ml 300 STE plus 0.04% w/v NaN_3

16. The material is now ready to be packed into a column.

17. Estimate the yield of oligonucleotide bound to the Sepharose by counting the radioactivity and compare with the results obtained at step 8. Usually between 20 and 40% of the DNA is bound to the column.

Protocol 5. Affinity chromatography

1. Pack a 3–5 ml column with the specific binding oligonucleotide of interest linked to Sepharose 4B. Pack an equivalent sized pre-column with a non-binding oligonucleotide linked to Sepharose 4B. Both columns should be 1 cm diameter and washed with 300 STE.

2. Connect the columns in tandem and wash with N200[a]-buffer at a flow rate of 0.2 ml/min at 4°C.

3. Thaw the DNA-binding protein sample and add the solution of the non-specific DNA competitor slowly with careful mixing. Leave on ice for 1 h to allow any precipitate to form.

4. Centrifuge at 40 000 × g for 6 min.

5. Pump the supernatant on to the affinity columns.

6. Wash the columns with N200 buffer until the absorbance at 280 nm returns to background.

7. Disconnect the pre-column and keep for later elution.

8. Wash the specific column with N200 buffer and then elute with a 60 ml linear salt gradient between 0.2 and 1.0 M NaCl (buffers N200 and N1000),[a] and collect 2 ml fractions.

Protocol 5. *Continued*

 9. Take aliquots of each fraction for band-shift analysis and SDS-PAGE. Store fractions on ice until these analyses are complete.

10. If required, elute pre-column with a similar gradient of N200 vs. N1000.

11. Wash both columns with 2 M NaCl TE followed by 300 mM NaCl TE, 0.04% NaN$_3$ and store at +4°C.

12. Protein fractions containing the DNA-binding activity may be:

 (a) stored (step 13)

 (b) recycled (step 14)

 (c) finally purified by reverse-phase chromatography (see below)

13. Dialyse active samples against 10 volumes of N100/50 buffer, which is N100 buffer containing 50% (v/v) glycerol.[b] After 4 h the sample can be centrifuged out of the dialysis bag and then stored at −20°C. Do NOT store at −70°C because this results in a large loss of activity, especially with these dilute protein samples.

14. Active samples from step 13 may be reapplied to the affinity column. Normally one would only use the specific column with a small amount of competitor, i.e. 1:1 ratio of competitor DNA to protein. So repeat steps 2–12 omitting steps 7 and 10.

[a] See *Table 1.*
[b] 50% v/v glycerol = 62.5% w/v.

6. Reverse-phase chromatography

Ideally, after the affinity chromatography there should be one band on an SDS gel in the fractions that contain all the DNA-binding activity. This is not usually achieved in one cycle of affinity chromatography. One can recycle the protein through the affinity column, in which case the amount of competitor DNA should be reduced. The yields of protein are usually quite high; in principle, therefore, this procedure can be continued until the protein is pure. At any stage when there is a simple mixture of proteins (say five bands) several methods (see Chapter 3) can be used to identify which protein is the one of interest: (a) renaturation of protein eluted from SDS gel slices (12); (b) photoaffinity cross-linking to labelled DNA (2); or (c) by two-dimensional electrophoresis of the band-shift gel (1). Once the protein has been unambiguously identified on the SDS-polyacrylamide gel, one is no longer reliant on non-denaturing forms of chromatography, and the high resolution of preparative electrophoresis and reverse-phase chromatography can now be used. Electrophoresis requires that the protein eluted from the affinity columns should be concentrated into a small volume of low-salt loading buffer.

Unacceptable losses usually occur on concentration and dialysis of dilute solutions. Some of the buffer components such as non-ionic detergents may also concentrate with the protein and interfere with the electrophoresis.

Reverse-phase chromatography, on the other hand, will remove the detergents and other buffer components, remove other protein contaminants, and concentrate the protein into a volatile buffer. *Figure 3* shows an elution profile of protein from the previous CTCF affinity chromatograph loaded on to a C_1/C_8 Pharmacia Pro RPC column and eluted with an acetonitrile gradient in 0.08% trifluoroacetic acid (TFA). The sample was diluted with TFA and then loaded by three injections using a 2 ml loop. As can be seen, the vast majority of the optical density does not bind to the column. Most of the other major peaks do not contain protein and are due to detergent and other buffer components used in the previous steps. The 130 kDa protein elutes at 35% acetonitrile in a volume of 0.7–1.4 ml. This protein is then available for direct sequence analysis and chemical or enzymatic cleavage.

Protocol 6. Reverse-phase chromatography

You will need:

- Apparatus: HPLC with two pumps, large (2 ml) loop, and gradient programmer, and a detector set at 214 nm. (The Beckman System Gold was used in *Figure 3*.)
- Column: Pharmacia Pro-RPC
- Solvent: A, 0.08% (v/v) TFA (Peirce); B, ~0.08% (v/v) TFA in 80% acetonitrile, 20% H_2O. The solvents should be made up and spurged with helium for 10 min and then the absorbance at 214 nm measured. Small aliquots of 10% TFA can be added to the A or B solvent with the lower absorbance until they balance.
- Gradient program:

Time (min)	% Solvent B	
0	5	Start fraction collector
3	5	
3–5	5–15	
5–45	15–70	
45	70	Stop fraction collector
45–50	70–100	
55–60	100–5	
60	5	
70	5	Stop flow

1. Connect the Pro-RPC column to the HPLC. The 5/2 (5 mm diameter × 2 cm) column is used at a flow rate of 0.7 ml/min. Up to 4 ml of detergent-containing buffers have been loaded on to this column with good results.

95

Protocol 6. *Continued*

The 10/10 (10 mm × 10 cm) column is used for 10–20 ml of sample and can be run at 2 ml/min.

2. Pre-wash the column with alternatively 100% solvent B and 100% solvent A a few times until the baseline stabilizes, and then return to the loading condition (5% solvent B).

3. Return the HPLC to manual control mode so that multiple loading can be carried out.

4. Remove one 2 ml protein sample from ice and hand-warm it to room temperature. Add 20 μl 10% (v/v) TFA to the sample, load it into the loop and inject the column.

5. After 3 min, reload the loop with protein as in step 3 and inject the column once the absorbance has returned to zero.

6. Set up the gradient programme. Start the fraction collector, recorder, and gradient at the same time and collect 1 min fractions.

7. Dry down 40 μl of each sample between fractions 15 and 40 using vacuum centrifugation (Savant Speed Vac concentrator). Redissolve in SDS sample solvent and analyse by SDS-PAGE, silver-staining the gel.[a]

8. The remainder of the sample should be rapidly frozen in solid CO_2 and stored at −70°C.

9. At the end of run, wash out the loop with solvent A and then wash the column and HPLC with methanol.

[a] Note some proteins do not stain well with silver. In these cases, gels may be electroblotted on to PVDF membranes such as Immobilon-P (Millipore) and stained with the gold stain Aurodye (Janssen) (1).

---

7. Production and isolation of peptides

In order to clone the gene, one should aim to get several peptide sequences since not all oligonucleotide probes (based on peptide sequence) are suitable for screening libraries. Some reasons why an oligonucleotide may not be suitable are:

(a) if they contain a high GC content

(b) if they are very redundant

(c) if they contain cross-hybridizing sequences

(d) if they code for the N-terminal end of the protein and the 5′ end of the gene is not represented in the library

Also, with two or more sequences one has the option of using a polymerase chain reaction (PCR) approach to obtain a larger probe to screen the library (see Chapter 6).

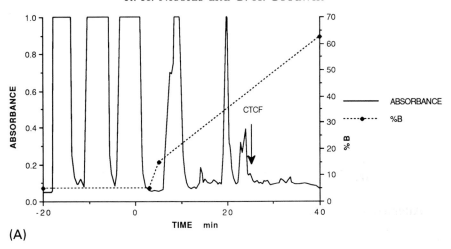

(A)

(B)

Figure 3. (A) Elution profile of the ProRPC reverse-phase chromatography of CTCF. Flow rate = 0.7 ml min^{-1}; λ = 214 nm; Buffer A = 0.08% TFA; Buffer B = 80% acetonitrile 0.08% TFA. Three 2 ml loadings were carried out at -20, -14, and -7 min and the gradient started at time zero. 1 min (0.8 ml) fractions were collected. (B) SDS-PAGE analysis of reverse-phase chromatography fractions. Lanes 1–20 correspond to fractions 21–40. 20 μl of each fraction was loaded. Lane C is 20 μl of affinity-purified CTCF and the marker lanes M were loaded with 100 ng and 25 ng per band. The Excel gel was stained with silver by the Pharmacia standard protocol.

The rather sparing quantities of the pure protein that are usually produced mean that there is very little opportunity to optimize the production and purification of the peptides. A test protein (e.g. bovine serum albumin) can be used but the peptide cleavage pattern of each protein is unique so one usually has to just 'go for it'. We have used two methods for separating peptides after cleavage: microbore reverse–phase chromatography and polyacrylamide gel electrophoresis followed by protein blotting (13). Both these methods will give sufficient peptides for sequencing using a minimum of 100 pmol of protein. There will always be losses associated with fractionating the peptides following cleavage, and not all the peptide immobilized on the support in the sequencer is chemically available. With 100 pmol of protein we obtained 5–10 pmol initial yield of amino acids at the start of a sequencing run, and a sequence of up to 20 amino acids can be obtained, depending on the nature and purity of the peptides. To obtain reliable sequences it is essential that sufficient pure protein is accumulated before committing it to cleavage and a sequence run.

The two choices that have to be made are the type of cleavage and the type of purification. Cleavage with CNBr is commonly used, because it is very specific for methionine which is a rare amino acid ($\sim$2%) in most proteins. This generates few but large peptides, which are best separated by SDS-polyacrylamide gel electrophoresis (SDS-PAGE). This method is especially useful for large proteins. Enzymatic digestions normally cut more frequently and this can make it more difficult to purify peptides in one reverse-phase chromatography step. However, we have used the enzyme Glu-C, which cleaves the C-terminal of glutamic and aspartic acid, to successfully generate peptides from GATA1 and have purified the peptides by reverse-phase chromatography.

Microsequencing is a field of its own and it is beyond the scope of this chapter to describe it in great detail. There are two good guides to micro-sequencing and related techniques (13, 14) and the second gives some recent developments using microbore columns for chromatographies other than reverse-phase chromatography. The sequence analysis of CTCF and GATA1 was carried out on an Applied Biosystems gas-phase sequentor 120 A with a 900 A data-handling system. Applied Biosystems also provide useful Application Notes.

Protocol 7. Preparation of peptides by CNBr cleavage and purification by reverse-phase chromatography

1. Concentrate purified protein from the preceding reverse-phase chromatography to 200 μl by evaporation using a Speed-Vac (Savant).[a] Do not evaporate to dryness.

2. Add 0.4 ml formic acid (AR), followed by 0.2 ml of 25 mg/ml CNBr in 70% v/v formic acid.

3. Flush with nitrogen and incubate in the dark at 25°C for 6 h.

4. Add a further 0.2 ml CNBr solution and incubate overnight.

5. Dry down sample in a speed-Vac with NaOH in the trap.

6. Add 100 μl of HPLC H$_2$O and dry down again. Continue to step 7 for reverse-phase chromatography or to *Protocol 8* for SDS-PAGE separation of peptides.

7. Redissolve sample in 100 μl 8 M guanidine–HCl, 20 mM Hepes (pH 7.9), 100 mM DTT.

8. Centrifuge sample and load on to microbore reverse-phase column using the following conditions:

- Brownlee C_{18} RP-300 column (2.1 × 100 mm)
- Solvent: A, 0.08% TFA; B, ~0.08% TFA, 80% acetonitrile
- HPLC syringe pump designed for microbore work, programmable
- UV detector: λ = 214, or variable wavelength
- Flow rate: 100 μl/min.
- Fraction collector: Collect 1 min fractions or collect peaks manually. (Remember to determine the time delay from the detector to the fraction collector.)
- Gradient program:

Time (min)	% Solvent B	
0	5	Start
0–5	5	
5–55	70	
55–60	100	
70–75	5	
90	5	Stop

9. Freeze all peak fractions rapidly and store at −70°. Losses tend to increase with storage, so if possible analyse the best peaks immediately.

[a] For enzyme cleavage, neutralize the protein sample from the preceding reverse-phase chromatography by adding Hepes to 20 mM. Evaporate off 50% of the solvent; this removes most of the acetonitrile. Check the pH and add enzyme. For these dilute solutions, a ratio of 1 μg sequence-grade enzyme (Boehringer) to 10 μg protein was used. Both Glu-C and trypsin will work well in 10% acetonitrile. After digestion, reduce with 100 mM DTT and continue to step 8.

Protocol 8. Purification of peptides by SDS-PAGE and blotting on to PVDF membrane

1. Make up a 0.75 mm thick SDS-polyacrylamide (Laemmli) gel with 15–20% polyacrylamide, depending on the sizes of the peptides in the mixture. Use highly purified acrylamide, ultra-clean apparatus, and HPLC-grade water for the buffers. Store the gel overnight.

2. Redissolve peptide sample in 20–40 μl SDS gel-loading buffer (2% w/v SDS, 62.5 mM Tris–HCl (pH 6.9), 10% (v/v) glycerol, 100 mM DTT) and heat to 90°C for 5 min.

3. Load gel and run as normal with markers but include 100 μm thioglycolic acid in the top reservoir buffer.

4. At the end of the electrophoresis, place the gel in transfer buffer (10 mM CAPS (pH 11.0), 10% methanol) for 5 min.

Protocol 8. *Continued*

5. At the same time wet three PVDF (Applied Biosystems) membranes in methanol and then soak in transfer buffer.

6. Make up the transfer sandwich with Whatman 3MM paper (soaked in transfer buffer) at the anode side, followed by two membranes, the gel, a protecting back-membrane of PVDF and more sheets of Whatman 3MM paper.

7. Place the sandwich in the apparatus and start the transfer. The time and current required depend on the size of the protein and design of the apparatus. The conditions should first be optimized for some test peptides in the same molecular weight range before committing the transcription factor peptide sample.

8. The backing membrane will have picked up some of the protein by passive blotting and may be stained with Aurodye for a permanent record.

9. Wash transfer membranes in HPLC water and then stain with 0.1% Coomassie blue R250 (PAGE blue 83, BDH) for 5 min and then destain with 50% (v/v) methanol, 10% (v/v) acetic acid and, finally, with methanol. Allow to dry. The detection limit is about 200 ng of peptide. Since this is the lower limit for sequencing, there is no need for more sensitive staining.

10. Cut bands out and place in gas-phase sequencer for sequencing. If protein bands are detected on the second transfer membrane, load them also into the sequencer. Bands can be stored at $-20°C$, if necessary.

---

8. Design of oligonucleotides for cDNA isolation

Typically, the sequencing of peptides will give you several sequences, 10–30 residues in length, which can be used to design oligonucleotides for PCR or library screening. At some positions, there may be doubt as to the amino acid, e.g. histidine and serine are sometimes not detected and tryptophan may be missed unless suitable precautions are taken.

Two approaches should be used to isolate a cDNA. The first is to use highly redundant oligonucleotides in the PCR reaction to amplify up a short fragment of DNA which (after cloning) can be used to screen a cDNA library for the full-length cDNA. This approach was first used to isolate a cDNA encoding the CREB transcription factor (15). The oligonucleotide primers can be 20–30 base pairs in length and can be up to 10^6-fold degenerate to accommodate all the codon possibilities, although we would advise restricting degeneracy to less than 10^5. The template is cDNA, preferably random primed. The PCR methodology is described in more detail in Chapter 6. The second strategy is the use of labelled oligonucleotides to screen the libraries. The problem here is that, unlike the PCR technique, it is difficult to use fully degenerate probes. If the complexity of probe is very high and only a small percentage of the oligonucleotides in the mixture are sufficiently homologous to the target sequence to hybridize, then the concentration of hybridizing probe is not sufficiently high to drive the annealing kinetics at a reasonable rate. Also, a large percentage of the radioactivity will be on incorrect oligonucleotides and this will contribute to high backgrounds on the filters. In practice, researchers have used either short mixed oligonucleotides to reduce degeneracy or single longer probes, guessing the third base at each codon or using inosine. Lathe (16) has advocated the use of a longer single probe, the sequence of which is optimized using codon usage considerations. The reader is well advised to consult this paper, not only for probe design but also for hybridization and washing conditions. Briefly, in order to screen a cDNA library of average complexity, the probe should be designed such that it has sufficient specificity to detect its target, and there should only be a 0.1 frequency of chance binding to spurious sequences in the library. This can be achieved with a 36-mer oligonucleotide, choosing the third base from a codon usage table and avoiding sequences encoding leucine, arginine, and serine. The average probability of correct choice of the third position is then 0.55 and gives an average predicted probe-target homology of 85%. Such a 36-mer is the minimum probe length required. In practice, one is likely to design a larger probe since there may be amino acids within the peptide chosen that were not identified or were not unambiguously identified, and there is a high chance that the chosen peptide has a serine, leucine, or arginine. Improved homology can be achieved by introducing a moderate level of complexity in the probe (giving a mixture of less than 128 oligonucleotides) by using two preferred bases at a few positions. Also, since deoxyinosine can base-pair with A and C (17), where the third base is likely to be T or G, this base can be incorporated. Inosine can also be used for the three bases of an unknown amino acid in the peptide sequence. With such considerations in mind, the CTCF cDNA was isolated by designing several oligonucleotides which were end-labelled and first checked by probing a Northern blot. One of the probes detected a band of sufficiently long mRNA to encode the protein. The sequence of this oligonucleotide was:

ATG	GAG/A	GGC/A	GAG/A	GCT	GTG
M	E	G	E	A	V

GAG/A	GCC/T	ATT	GTG	GAG/A	GA
E	A	I	V	E	E

This oligonucleotide was end-labelled with polynucleotide kinase to a specific activity of more than 10^8 c.p.m./µg (*Protocol 9*) and used to probe a cDNA library. The most convenient procedure for screening high-complexity libraries ($\sim 10^6$ recombinant) is to plate out the phage at high density in large rectangular dishes and grow the phage until the plaques are less than 1 mm diameter. Duplicate filters are lifted and hybridized (24–48 h) in pairs, back to back, in sealed bags (*Protocol 10*) at a temperature 10–25 °C below the expected T_m of the hybrid. This can be calculated from the formula:

$$t_m = 102 - 820/l - 1.2 \, (100 - h)$$

where t_m is the melting temperature in $6 \times$ SSC, l is the length of the probe, and h is the percentage homology between probe and target.

The filters are washed with SSC buffers at increasing temperatures, auto-radiographing after each wash. Phage are extracted from positive plaques and rescreened through two more rounds. The cloned phages are then amplified and the DNA extracted. The inserts can be amplified and isolated by PCR and checked by Northern blot hybridization for correct size or tissue distribution, before subcloning and sequencing.

Protocol 9. End-labelling of oligonucleotide

1. Mix the following on ice:
   - 2 μl buffer K (*Protocol 4*)
   - 2 μl oligonucleotide (360 pmol, ~0.7 μg)
   - 15 μl [γ-^{32}P]ATP 10 mCi/ml (5000 Ci/mmol)
   - 1 μl (10 IU) T_4 polynucleotide kinase (New England BioLabs)
2. Incubate at 37 °C for 45 min.
3. Remove unincorporated nucleotides by the spinning-column method using DNA-grade G-25 Sepharose or NAP-5 column equilibrated in 10 mM Tris (pH 7.6), 1 mM EDTA. The specific activity of the probe should be $>10^8$ c.p.m./μg.

Protocol 10. Hybridization of oligonucleotide to library filters

1. The cDNA library is plated out at a density of approximately 5×10^5 plaques per 400 cm^2 in $24.5 \times 24.5 \times 2$ cm dishes (Nunc). The plaques are lifted in duplicate on to 20×20 cm nitrocellulose filters (Amersham), alkali denatured, and neutralized as described by Maniatis *et al.* (12).
2. The filters are cross-linked with UV light using the UV Stratalinker 1800 (Stratagene). Use the auto cross-linking conditions (1.2×10^5 μJ).
3. The filters are pre-hybridized overnight in sealed bags; two duplicate filters back to back per bag. Pre-hybridization buffer is $6 \times$ SSC,[a] 0.5% SDS, 100 μg/ml sonicated denatured salmon sperm DNA, $5 \times$ Denhardt's solution. The temperature of pre-hybridization is the same as that chosen for hybridization (40–55 °C).

4. The solutions are replacd with 20–30 ml per bag hybridization buffer containing the labelled probe (10^6–10^7 c.p.m./ml). Hybridization buffer is 6 × SSC, 0.1% SDS. The temperature of hybridization should be 10–25°C below the T_m. Hybridization is carried out for 24–48 h, depending on the complexity of the probe. The filters are then washed four times for 5 min each in 6 × SSC at room temperature, then at steps of increasing temperatures until the T_m is reached, autoradiographing the filters between washes and exposing to the X-ray film for approximately 4 h at −70°C with an intensifying screen. Each high-stringent wash should be for 1–5 min.

[a] 20 × SSC is 3 M NaCl, 30 mM sodium citrate.

Acknowledgements

The work described in this chapter was supported by the Cancer Research Campaign. The authors thank A. Carne for peptide sequencing, and N. Perkins, C. Heath, and E. Klenova for their contributions in developing the techniques described in this chapter.

References

1. Perkins, N. D., Nicolas, R. H., Plumb, M. A., and Goodwin, G. H. (1989). *Nucleic Acids Res.*, **17**, 1299.
2. Wu, C., Wilson, S., Walker, B., Dawid, I., Paisley, T., Zimarino, V., and Ueda, H. (1987). *Science*, **238**, 1247.
3. Wang, J., Nishigama, K., Araki, K., Kitamura, D., and Watanabe, T. (1987). *Nucleic Acids Res.*, **15**, 10105.
4. Lobanenkov, V. V., Nicolas, R. H., Adler, V. V., Paterson, H., Klenova, E. M., Polotskaja, A. V., and Goodwin, G. H. (1990). *Oncogene*, **5**, 1743.
5. Evans, T. and Felsenfeld, G. (1989). *Cell*, **58**, 877.
6. Dignam, J. D., Lebovitz, R. M., and Roeder, R. G. (1983). *Nucleic Acids Res.*, **11**, 1475.
7. Kadonaga, S. T. and Tjian, R. (1986). *Proc. Natl. Acad. Sci. (USA)*, **83**, 5889.
8. Kadonaga, J. T. (1991). In *Methods in Enzymology*, Vol. 208 (ed. R. T. Sauer), pp. 10–23. Academic Press, London.
9. Ferguson, K. A. (1964). *Metabolism*, **266**, 3052.
10. May, G. E., Sutton, C., and Gould, H. (1991). *J. Biol. Chem.*, **266**, 3052.
11. Maniatis, T., Fritsch, E. F., and Sambrook, J. (ed.) (1982). *Molecular Cloning, A Laboratory Manual*. Cold Spring Harbor Press, Cold Spring Harbor, NY.
12. Hager, D. A. and Burgess, R. R. (1980). *Anal. Biochem.*, **109**, 76.
13. Matsudaira, P. T. (ed.) (1989). *A Practical Guide to Protein and Peptide Purification for Microsequencing*. Academic Press, London.
14. Nice, E. C. (1990). *Nature Lond.*, **348**, 462.

15. Gonzalez, G., Yamamoto, K. K., Fischer, W. H., Karr, D., Menzel, P., Biggs, W., Vale, W. W., and Montminy, M. R. (1989). *Nature*, **337**, 749.
16. Lathe, R. (1985). *J. Mol. Biol.*, **183,** 1.
17. Kawase, Y., Iwai, S., Inoue, H., Miura, K., and Ohtsuka, E. (1986). *Nucleic Acids Res.*, **14,** 7727.

5

Cloning transcription factors from a cDNA expression library

IAN G. COWELL and HELEN C. HURST

1. Introduction

The initial chapters in this book describe methods of investigating the DNA-binding activity of a transcription factor (see Chapters 1 and 2) and how to use these properties to purify the protein (see Chapter 4). However, further studies concerning the functioning of the factor in transcription can be extremely limited unless a cDNA clone encoding the factor is isolated.

There are a number of ways of cloning DNA-binding factors, and the most suitable method will vary from case to case. If, for example, the factor of interest has been isolated in sufficient quantity and purity for reliable peptide sequence to be obtained, the peptide sequence data may be used to design oligonucleotide probes to screen cDNA libraries by hybridization (see Chapter 4). However, a number of alternative approaches utilize cDNA expression libraries, where specific clones are detected through some property of the encoded polypeptide, such as its DNA-binding activity or immunological properties. Commonly, cDNA expression libraries are constructed using bacteriophage vectors such as λ gt11, where the cloned cDNA is expressed as a fusion protein with β-galactosidase upon addition of an inducer (1). The advantages of such a vector include rapid screening of large numbers of recombinants and large library size. Pertinent points concerning the handling of bacteriophage libraries are included in Section 2.

As alluded to above, one method of screening bacteriophage expression libraries is the use of DNA probes comprising the binding site of the factor in question. This approach has been particularly successful in the cloning of members of the leucine zipper family of factors (2–5), and has also been used for other classes of DNA-binding protein (6–11). The relative merits and shortcomings of this approach and of immunological screening are discussed in Section 3.1, and protocols for each method appear later in the section.

2. Handling bacteriophage expression libraries

This topic is covered exhaustively in a previous edition of this series (1). However, the main points are presented here for completeness, and particular issues relevant to this type of screening are discussed.

2.1 Library selection

An obvious prerequisite is that the cDNA expression library should be generated from a cell line or tissue known to express the factor of interest in reasonable abundance. It is then the investigator's choice whether to make their own library or purchase one. In either event, the library should have a complexity of at least 10^6. In theory only one in six inserts will be in the correct orientation and frame to generate a β-galactosidase fusion protein. However, we have found in practice that proteins are often translated from an internal AUG codon. This can mean that the genuine initiator AUG may be used in cloned cDNAs complete at their 5' end: 5' untranslated regions often contain inframe stop codons, thus preventing readthrough from β-galactosidase.

If DNA-binding site probes are used for screening (see Section 3.2), a full-length cDNA is not usually required, but sufficient coding information to generate a complete DNA binding/dimerization domain is essential. As this domain may lie at the very N-terminus or C-terminus of the protein, a mixed library using random priming and oligo-dT priming is probably best. This is especially true if previous characterization of the factor indicates that it is quite large (say, >60 kDa). However, smaller proteins may also be difficult to clone if they have exceptionally long mRNAs. For example, the 46 kDa CREB protein is encoded in 1023 bases within a 7 kb mRNA.

The following protocols assume the use of a λ gt11 library, but libraries constructed in λ ZAP (Stratagene) can also be used by modifying the phage growth protocol to comply with the supplier's instructions. We have used a commercially available (Clontech) human placenta library in λ gt11 which we have found to have a high complexity, containing cDNA clones for a wide range of transcription factors. However, few of these clones are full length, apparently due to incomplete *Eco*RI methylation of the nascent cDNA prior to cloning.

2.2 Library plating for screening

The home-made (preferably unamplified) or commercial library should first be titred on the bacterial expression strain Y1090 (Clontech). Make fresh plating cells by inoculating one colony into 20 ml of L-Broth containing 0.2% maltose. For all bacteriophage λ work use L-Broth containing 10 mM $MgSO_4$. Shake the culture vigorously until the stationary phase is reached, then pellet the cells and resuspend in half their original volume using SM buffer (see *Protocol 1*). Plating cells may be stored on ice until required (up to 24 h).

Once the titre has been ascertained, the library can be plated on 15 cm plates and protein expression can be induced for transfer to nitrocellulose membranes as detailed in *Protocol 1*.

Protocol 1. Library plating and replica lifts

You will need the following:

- Well-dried L-agar 15 and 9 cm diameter plates
- Autoclaved top agarose: 0.7 g agarose in 100 ml L-Broth. Melt in microwave and maintain at 45°C for 20 min before use
- SM buffer to dilute phage: 100 mM NaCl, 10 mM MgSO$_4$, 50 mM Tris–HCl (pH 7.5), 0.01% gelatin. Autoclave
- Fresh Y1090 plating cells
- Nitrocellulose filters (Schleicher and Schuell). 132 and 82 mm circles
- Isopropyl β-D-thiogalactopyranoside (IPTG). 1 M stock in water. Store at −20°C
- Wash buffer: 50 mM NaCl, 10 mM Tris–HCl (pH 7.5), 1 mM EDTA, 1 mM dithiothreitol (DTT), and 0.05% lauryl dimethylamide oxide (LDAO; Calbiochem)

1. Plate out at a density of 5×10^4 bacteriophage per 15 cm plate. For each plate, dilute phage in 200 μl of SM buffer and add 600 μl of fresh Y1090 plating cells. Leave to infect for 15 min at 37°C then add 7.5 ml top agarose and pour on to pre-warmed plates. Aim to analyse 10–20 15 cm plates in the primary screen.

2. When all plates are set, transfer to a 42°C incubator until plaques are just visible as pin pricks on the bacterial lawn. This takes 3–4 h.

3. Soak one nitrocellulose filter per plate in 10 mM IPTG and pat dry. Working quickly so that the agar temperature does not drop below 37°C, take one plate at a time and place a still damp numbered filter on the surface, then transfer the plate to a 37°C incubator. Incubate for 1–2 h.

4. Pierce the filter and agar with three to four orientation holes using a syringe needle dipped in India ink. Carefully remove the filter with forceps and transfer to a sandwich box containing 500 ml of wash buffer. Place a second IPTG-soaked filter on each plate in turn and incubate at 37°C for a further 2–3 h.

5. With a permanent marker pen, mark the second set of filters with dots over the syringe holes in the agar. A light box makes this easier. Remove these filters to the wash buffer. All filters should be washed for 5–10 min to remove adherent bits of agarose and so reduce background. The filters are now ready for blocking (see Sections 3.2 or 3.3, depending on screening method to be employed).

2.3 Plaque purification

The screening process is described in the next section. Once positives are detected on duplicate filters, corresponding plugs of agar should be picked from the plate for further rounds of screening and for phage purification. Using the orientation holes in the agar, align the plate with the developed autoradiograph and use the wide end of a Pasteur pipette to pierce the agar. Transfer the resulting agar plug into 1 ml of SM buffer containing a drop of chloroform in a bijou bottle. The bottles can be shaken gently to speed phage elution (2 h to overnight). Ideally, the eluted phage should be titred before plating for second-round screens. Plate out the titred, eluted phage as described in *Protocol 1*, but this time at a density of 500–1000 plaques on a 9 cm plate. It is possible to use less-dense plates, but it is more reassuring if second-round screens contain several positive signals per plate! Duplicate lifts are not always necessary for second and further screens.

Screen these second-round filters and pick positive areas of agar using the narrow end of a Pasteur pipette, again into 1 ml of SM plus a drop of chloroform. The phage eluted from these plugs should be titred and replated at low density on 9 cm plates for third-round screens. After this stage, it should be possible to pick a pure clean phage plaque, however, it is wise to perform a fourth-round screen to check for purity as phage particles do diffuse readily on agar plates.

3. Screening methods

3.1 Introduction

Once the filters carrying the expression library have been generated, they may be probed using one of the methods described in this section. Two approaches will be considered in detail: the use of DNA-binding-site probes and immunological screening. Each of these methods has its own merits and drawbacks; the use of DNA-binding-site probes has the advantage that little prior characterization of the factor or factors of interest is required before embarking on library screening, and it is not necessary to obtain highly purified fractions of the protein. This feature can be useful if it is simply desired to clone factors that bind a particular sequence, such as a defined regulatory element in a promoter, when little detail is known about these factors. However, this lack of specificity can be a problem if one particular factor is to be cloned, since it is frequently the case that a family of factors, rather than a single species, bind any given regulatory element (see Section 3.2.2). However, as discussed in Section 3.2.1, this problem can often be minimized by careful selection of probe sequence. Some other limits to the effectiveness of binding site screening are outlined below.

(a) If post-translational modification such as phosphorylation or specific proteolytic cleavage is required for efficient DNA-binding activity, then the factor expressed in *E. coli* may not bind DNA with high affinity. Careful selection of binding and washing conditions may circumvent this problem (see Section 3.2).

(b) Problems might arise from poor solubility or protein folding of factors expressed in *E. coli*. However, this need not be intractable as described in Section 3.2.5.

(c) A third problem with binding-site screening arises if the factor of interest binds DNA efficiently only as a heterodimeric or heteromeric complex, since only a single species can be expressed from any one clone. Similarly, this method is of no value in the cloning of components of a multisubunit transcription factor complex if they do not stably bind DNA themselves in the absence of the other components.

Immunological screening has the advantage that DNA-binding activity is not required from the expressed product of the cDNA clone and so the potential problems discussed above do not arise. However, there are potential drawbacks associated with this method as well. Firstly, prior purification of the protein is required in order to raise an antibody for screening. Secondly, as described in Section 3.3.3, unexpected results can arise due to lack of specificity or cross-reactions with the antibody. A third drawback of immunological screening when compared with screening with DNA-binding-site probes concerns the ease of analysing the clones that are obtained. Since a universal feature of transcription factors is DNA binding, the first feature to test when checking the identity of a clone is its DNA-binding activity and specificity. This is straightforward if DNA-binding activity was the feature used to select the clone (see Section 4), but this is not necessarily the case for immunological screening, since the clones that are obtained may be partial and will not always contain the sequences required for DNA binding.

3.2 Screening with DNA-binding-site probes

Before embarking on library screening, it is advisable to determine the optimum conditions for binding in terms of specific probe sequence, non-specific competitor DNA, and binding buffer composition. These points are considered in Section 3.2.1. A procedure for screening with binding site probes is given in Section 3.2.4, and a variation of the protocol appears in the following section. Further notes and fault-finding hints are given in Section 3.2.6.

3.2.1 Selection of probe sequence and binding conditions

The use of appropriate binding-site probes is obviously essential for successful cDNA library screening. In general, the binding site that is used will be one that is known from gel retardation, or other experiments, to be bound by the factor of interest. Probes containing multiple binding-sites give good signals and may take the form of a single, long oligonucleotide containing several sites for the factor of interest, a restriction fragment containing multiple cloned binding-site oligonucleotides or alternatively, and in our hands preferably, a concatenated binding-site-containing oligonucleotide. The rest of this section assumes the use of this third type of probe. For the construction of concatenated probes, the specific binding sequence for the factor of interest should be contained in a 20- to 25-mer double-stranded oligonucleotide with free 3'-hydroxyl groups to facilitate labelling with γ-[^{32}P]ATP and compatible 'sticky ends' to allow ligation. The sequence of the oligonucleotide should be chosen, if possible, such that it does not contain binding sites for DNA-binding factors other than the ones of interest. The specificity of a new oligonucleotide sequence can easily be tested by gel retardation analysis (see Chapter 1). However, another consideration when designing a probe is the nature of the new sequences that are generated at the ligated junctions of the concatenated probe. The generation of binding sites for other factors at these positions should also be avoided as much as possible in order to simplify the screening procedure.

Gel retardation analysis can also be used to determine the best non-specific competitor DNA to use. We have found poly(dA)·poly(dT) to be suitable, but poly(dI)·poly(dC) or herring sperm DNA may be more suitable for some factors.

We found TNE-50 (see *Protocol 2*) to be a suitable binding and washing buffer. Its low salt concentration encourages the binding of bacterially synthesized proteins which may bind DNA relatively poorly due to lack of post-translational modifications etc. Binding activity in TNE-50 should preferably be tested in advance for the factor of interest by either gel retardation or Southwestern blot analysis.

3.2.2 Southwestern blotting

As the name implies, Southwestern blotting is a variation of the traditional Western blotting technique. Following transfer on to a nitrocellulose membrane, electrophoresed DNA-binding proteins are detected with a labelled DNA probe (rather than by immunological means). The method therefore embodies the technique used to screen the library. A protocol for Southwestern blotting is given below. The sample material could be whole-cell extract, nuclear extract, or partially purified DNA-binding protein.

Protocol 2. Southwestern blotting

You will need the following:

- TNE-50: 10 mM Tris (pH 7.5), 50 ml NaCl, 1 mM EDTA, 1 mM DTT
- SW-block: 2.5% (w/v) dried milk powder, 25 mM Hepes (pH 8.0), 1 mM DTT, 10% (v/v) glycerol, 50 mM NaCl, 0.05% (v/v) LDAO (Calbiochem), 1 mM EDTA

1. Subject sample to SDS-polyacrylamide gel electrophoresis and transfer on to a nitrocellulose membrane using standard Western blotting techniques.

2. After transfer, gently wash the membrane in TNE-50 and place in a tray or sandwich box containing enough SW-block to completely immerse the membrane. Block overnight at 4°C.

3. Remove the filter from the SW-block solution and wash it briefly in TNE-50.

4. Immerse the membrane in a probe mixture consisting of:
   - 10 μg/ml non-specific competitor DNA
   - 2×10^6 c.p.m./ml labelled DNA probe (see Section 3.2.3)
   - TNE-50

   Incubate for 1–2 h at room temperature in a small tray or sandwich box, or alternatively in a sealed plastic bag.

5. Remove the membrane from the probe solution and wash in 100–200 ml of TNE-50 for 5–10 min.

6. Repeat this washing two to three times or until the radioactive level of the membrane no longer falls appreciably between washes.

7. Blot the filter dry, wrap in cling film, and expose to X-ray film.

A successful Southwestern blot would reveal a single hybridizing band with a mobility consistent with the molecular weight of the factor of interest, as determined from other studies such as UV cross-linking (see Chapter 3). Two important pieces of information that are revealed by Southwestern blotting are:

(a) If the factor of interest is able to bind the probe DNA after electrophoretic separation, it must be able to bind as a single species, i.e. a heterodimeric or heteromeric complex is not required for binding. This is clearly important when screening a library with binding-site probes.

(b) If more than one band is present on the autoradiograph after Southwestern blotting, the probable explanation is that the probe sequence is bound by a family of factors. In this event, it may be advisable to try and determine a more selective probe or be prepared to clone and characterize several members of a family of factors.

3.2.3 Preparation of oligonucleotide probe

Described below is the preparation of concatenated binding site probes.

i. Annealing

Having decided on the sequence of the oligonucleotide probe to use, it is important to obtain efficient annealing of the two synthesized strands. Take

5 μg each of the upper and lower strand oligonucleotides and place in a 1.5 ml microcentrifuge tube. Add 10 μl of 10 × T4 kinase buffer (500 mM Tris (pH 8.0), 100 mM MgCl$_2$, 5 mM DTT, 1 mM spermidine). Make the volume up to 100 μl and place in a water bath at 90°C. Turn the water bath off and allow to cool to room temperature overnight. Store the double-stranded oligonucleotide at −20°C until required.

ii. *Probe labelling and ligation*
Double-stranded binding-site oligonucleotides are labelled with T4 poly-nucleotide kinase and [γ-^{32}P]ATP and then ligated with T4 DNA ligase. The concatenated probe should then be used within 24 h.

Protocol 3. Probe preparation

You will need:

- 10 × polynucleotide kinase buffer: 500 mM Tris (pH 8.0), 100 mM MgCl$_2$, 5 mM DTT, 1 mM spermidine. Make up this solution freshly for maximum labelling efficiency.
- TE: 10 mM Tris (pH 8.0), 0.1 mM EDTA

1. Mix in a microcentrifuge tube:
   - double-stranded oligonucleotide (0.1 μg/μl)   4 μl
   - 10 × polynucleotide kinase buffer   2 μl
   - [γ-^{32}P]ATP (10 μCi/μl; 5000 Ci/mmole)   5 μl
   - T4 polynucleotide kinase (10 IU/μl)   2 μl
   - H$_2$O   7 μl

2. Incubate at 37°C for 60 min.

3. Add 80 μl of TE and separate the labelled oligonucleotide from the excess ATP by spun-column chromatography (we use a 1 ml syringe plugged with polyallomer wool and filled with Sephadex G-50M).

4. Check the incorporation of radioisotope by counting 1 μl of the spun-column eluate by liquid scintillation counting. (The labelling reaction should yield 1–2 × 10^8 c.p.m. in total.)

5. To the labelled oligonucleotide (approximately 100 μl) add:
   - 10 × T4 DNA ligase buffer (New England Biolabs)   11 μl
   - 10 mM ATP   2 μl
   - T4 DNA ligase (New England Biolabs, 10 IU/μl)   2 μl

6. Ligate overnight at 15–18°C.

3.2.4 Library screening

Filter lifts (from Section 2) must be blocked before screening with labelled probe. For 10–20 filters, place 500 ml of SW-block (see Section 3.2.2) in a tray or large sandwich box. Immerse the filters in the blocking solution one at a time, ensuring that both surfaces of each filter come into contact with the solution. Block overnight at 4°C. Filters may then be probed as described in *Protocol 4*.

Protocol 4. Filter screening

1. Make up the probe mixture by adding the concatenated probe (*Protocol 3*) to 100 ml of TNE-50 (*Protocol 2*) containing non-specific competitor DNA at 10 μg/ml. The final radioactive concentration should be $5 \times 10^5 - 1 \times 10^6$ c.p.m./ml.

2. Place 50 ml of the probe solution into each of two 150 mm Petri dishes.

3. Lay filters one at a time in the probe mixture, protein side up, placing only five to six filters in each dish.[a]

4. Incubate for 1 h at room temperature, using a slowly moving orbital platform shaker to keep the filters floating freely.

5. Remove the filters to a large tray or sandwich box and wash three to four times for 5–10 min each time in 200–300 ml of TNE-50 at room temperature, making sure that the filters remain free in the solution and do not stick together. Washing can be performed on an orbital platform shaker with fairly vigorous movement.[b]

6. Blot the filters dry and expose to X-ray film including fluorescent or radioactive markers to facilitate orientation of the filters to the developed film.

[a] Process the filters in batches if necessary, re-using the probe.
[b] The probe mixture may be retained and kept at −20°C for up to one week and may be used for second-round screens after adding fresh DTT immediately before use.

Using Kodak X-omat AR film and an intensifying screen, an overnight exposure at −70°C should be sufficient to detect clones giving strong signals. However, if the background allows, a second, longer exposure (1 week) is also recommended in order to detect clones giving weaker signals. *Figure 1* shows the result of typical first- and second-round screens after a 24 h exposure. In this particular case, a concatenated ATF site probe was used to screen for members of the CREB/ATF family.

The hybridization conditions described above are given as a guide and represent conditions that we have found to work well. Longer incubation times at a lower temperature may result in a better signal from a poorly

3.3 Immunological screening

If it is possible to purify reasonable amounts of the factor of interest, then two routes to cloning the cognate gene are open. These are either to derive peptide sequence information and thence screen a library with redundant oligonucleotides, as described in Chapter 4, or to use the purified protein to raise antibodies for use in screening expression libraries. In some cases, it may be possible to adopt both routes because peptide sequence information may also be used to make synthetic peptides that, once coupled to a suitable carrier, can be used to raise antibodies in rabbits.

3.3.1 Antibody considerations

The use of synthetic peptide to raise polyclonal antisera has the advantage that a large amount of immunogen and hence antiserum can be generated. In practice, however, peptide antisera often do not recognize the native protein. Consequently, antisera raised against the purified protein are usually preferable. If a small amount of protein has been purified, but the quantity is insufficient to ensure reliable peptide sequencing, then a good compromise may be to use the material to raise polyclonal antisera in mice or rats. The use of mice and rats has the advantage that only small amounts of antigen are required to activate the immune system and, additionally, if useful antisera are obtained, they can be used to generate monoclonal antibodies. The obvious disadvantage is that the yield of polyclonal serum is very small, and while it is possible to use monoclonal antibodies for screening expression libraries, polyclonal antisera are generally considered to give better results. One way round this may be to use a mixture of monoclonal antibodies in the screening process. The other problem is whether the purified factor will be sufficiently antigenic to produce an immune response in the first place, especially as nuclear factors are generally highly conserved across species. In practice, however, many factors have been used to raise good antisera. In *Protocol 6* we describe a method adapted for using small amounts of purified material which we have successfully used to raise mouse antisera against ATF-43 (12).

Protocol 6. Generating nuclear factor antibodies in mice

You will need:

- 2 or 3 F1 hybrid mice, e.g. Black 10 × Balbc. Pre-bleed to obtain pre-immune serum

- Affinity-purified factor. Each mouse must be injected at least twice and possibly three times with 50–200 ng of material, as estimated by gel retardation assay (see Chapter 1)

1. Precipitate 50–200 ng of purified protein by adding cold (−80°C) acetone to 80%. Resuspend in 300 μl of 0.9% saline. If the protein is thought not to precipitate well, then try to concentrate to approximately 100 ng in 300 μl of dialysis buffer (up to 10% glycerol, but no salt) using a centricon spin concentrator (Amicon).

2. To each 300 μl sample, add 90 μl of a sterile aqueous solution of 10% potassium alum (aluminium potassium sulphate) and mix.

3. Carefully add 12 μl of 4 M NaOH and mix well. A thick white precipitate forms, which consists of a complex of Al(OH)$_3$ and protein. This adjuvant/protein suspension can be subcutaneously injected into the mouse at multiple sites.

4. Repeat the sample preparation and injections 3 weeks after the primary injections. Test bleeds can be taken 2 weeks after the secondary injections. If further boosts are required, these can be given at 3- to 4-weekly intervals.

The easiest way to test antibody efficacy is to include 1–2 μl of serum in a gel retardation assay sample. Incubate for 2 h to overnight alongside control incubations containing pre-immune serum, before adding the probe, and then proceed as usual (see Chapter 1). If the antibody cross-reacts with the purified factor, then the normal protein–DNA complex will either be abolished or further retarded in its migration, depending on which protein domain(s) are recognized (see reference 12). Promising antisera can then be further tested by Western blotting. This is a good guide as to how well the antibody will recognize protein expressed from a bacterial expression library, because in both cases the antigen is presented on nitrocellulose filters in a partially denatured state. However, if the factor is normally post-translationally modified (e.g. heavily phosphorylated) then the antiserum may not recognize the same protein devoid of these modifications made in bacteria. One possibility is to treat the purified factor before using it as an immunogen—for example, phosphate may be removed using calf intestinal phosphatase immobilized on a column support (Scotlab).

Before using polyclonal antisera to screen an expression library, it is best to remove any contaminating antibodies to bacterial and phage proteins, which these preparations often contain. This will reduce the background in the primary screens. Concentrated bacterial/phage lysate can be obtained commercially and is often provided as part of a screening kit (e.g. Stratagene's *pico*blue immunoscreening kit). Incubate three to four strips of nitrocellulose in the lysate at room temperature, then rinse in phosphate-buffered saline (PBS) and air dry. The filters can then be used sequentially to pull out cross-reacting antibodies from polyclonal serum previously diluted 1:5 or 1:10 in PBST (PBS plus 0.05% Tween-20). This serum may represent the bleed out of one mouse, so care must be taken not to waste the sample. The cleaned-up serum can be stored at −20°C until needed.

3.3.2 Library screening

Wash the filter lifts from Section 2 (*Protocol 1*) in PBST and block overnight at 4°C in 500 ml of PBST supplemented with 1% bovine serum albumin (BSA; Sigma). Place 50 ml of a suitable dilution (1 in 10 to 1 in 25) of the cleaned up, diluted serum in blocking solution in a 15 cm Petri dish. Add batches of five to six filters one at a time so that they do not stick together, and incubate for 1 h at room temperature with orbital shaking. Wash the filters three times in PBST. The primary antibody can be stored at −20°C and re-used for subsequent screens. This re-use will also reduce the background.

Meanwhile, the filters are developed either using an enzyme-conjugated second antibody followed by a colour reaction, or ^{125}I-labelled protein A followed by autoradiography. Commercial kits are available to facilitate these steps. Either way, plaques giving positive signals should be picked and re-screened as described in Section 2.3

Once a pure preparation of a positive phage has been achieved, steps should be taken to establish whether this truly encodes the factor of interest. This is most easily done by preparing protein extract from a lysogen and testing this in a gel retardation assay (see Section 4). However, if the cDNA clone is incomplete, the DNA-binding domain may not be contained within the purified phage, and further screening steps using the initial clone as a hybridization probe may be required to obtain a full-length clone.

3.3.3 Potential problems with immunological screening

One particular problem with immunological screening is that it is possible to clone a protein completely distinct from the intended one. This arises when the material used to inject the animals to raise the antiserum is not totally free of contamination. A co-purifying protein may be a very minor impurity and may not be detected on silver-stained gels, but it may be very immunogenic. Consequently, the polyclonal serum will contain antibodies not only to the factor of interest, but also to this contaminant. We have found that the 68 kDa subunit of human Ku-antigen, which binds to the ends of dsDNA molecules, produces a very strong immunogenic response, even when the protein is present in undetectable amounts in highly purified factor preparations (A. Skinner & H. C. Hurst, unpublished data). Unfortunately, this can mean that the investigator ends up cloning a rogue contaminant rather than the factor of interest. Consequently, when picking plaques from primary screens for further purification, it is important to persevere with weak as well as strong signals. Of course, the accidental cloning of a co-purifying protein that influences the function of the factor of interest could be highly advantageous!

3.4 Other approaches to library screening and their relative merits

We described in Sections 3.3 and 3.2 methods for screening cDNA expression libraries immunologically and with DNA-binding site probes. A third method, which we have not discussed so far, involves screening of an expression library with a probe consisting of a labelled protein, possibly a subunit of a multicomponent transcription factor. The factor or subunit of interest is then detected by virtue of its protein–protein interaction with the labelled probe. This method is exemplified by MacGregor *et al.* (13) who screened a cDNA expression library with a biotinylated *c-jun* probe to detect other members of the leucine zipper family with which *c-jun* could dimerize. This screening strategy has value in the cloning of components of a transcription factor complex which alone bind DNA poorly or not at all and in the cloning of ancillary proteins with which transcription factors interact. A shortfall of this method is that only strong protein–protein interactions such as occur between leucine zipper factors are likely to result in a stable enough interaction for library screening.

Expression of cDNA clones in eukaryotic cells provides an alternative to screening bacteriophage libraries. In this approach, cDNA libraries are constructed using a high-level transient expression vector plasmid. COS cells or other cells allowing high-level expression are transfected with pools of cDNA-containing plasmids. Nuclear extracts are then prepared from transfected cells and assayed by gel retardation analysis for the presence of a binding activity corresponding to the factor of interest. Plasmid DNA is recovered from transfected cells expressing the expected DNA-binding activity, and recovered DNA is used for successive rounds of selection, until a pure clone is obtained. The use of this approach is exemplified by Tsai *et al.* (14) who cloned the erythroid specific factor GF-1 (GATA) by transient expression in COS cells. This approach is most appropriate where the DNA-binding activity of interest is highly tissue-specific or inducible, and it has the advantage that cloned factors expressed in cell culture cells are far more likely to receive appropriate post-translational modification than bacterially synthesized factors. In addition, this procedure allows rapid comparison between the cloned factor and the endogenous protein in terms of DNA-binding specificity and mobility in gel retardation analysis. However, as with screening bacteriophage expression libraries with binding-site probes, this strategy is only useful if the binding activity of the factor is contained in a single protein species.

Some workers have designed complementation approaches to obtain cDNA clones for particular transcription factors. For example, Becker *et al.* (15) cloned the CCAAT binding factor CP1 by complementation of a yeast HAP2 mutant. HAP2 is the yeast homologue of CP1; HAP2 mutants are viable but suffer from a respiratory deficiency. Transformants expressing a clone capable of replacing HAP2 were selected for their ability to grow on minimal agar. Another example of cloning by complementation is provided by Zhou and Thiele (16) who cloned the *Candida glabrata* metal-activated transcription factor in *Saccharomyces cerevisiae*.

Cloning by complementation is clearly not a universally applicable means of cloning transcription factors, and the strategy and detailed methodology will be different in each case. However, it has the advantage that the clone is selected on the relatively stringent basis of expressing a particular function.

4. Proving the identity of the factor

When cloning transcription factors from λ phage cDNA expression libraries or by other means, it is usually desirable to prove as rapidly as possible the identity of the cloned factors. The criteria that are readily tested are listed below:

(a) DNA-binding specificity

(b) immunological properties

(c) amino acid sequence

(d) transcriptional activating/repressing potential

By use of the λ gt11 system, it is possible to generate bacterial colonies carrying an integrated copy of the purified factor-encoding phage (a lysogen). The advantage of this is that a liquid culture can then be grown up and induced to synthesize the cloned factor (usually, but not necessarily, as a fusion protein with β-galactosidase) within the cells. Crude protein extracts prepared from these cells will be enriched in the cloned recombinant protein and can be assayed in much the same way as crude nuclear extracts prepared from eucaryotic cells.

A detailed method for generating a lysogen using the bacterial strain Y1089 (Clontech) is given in *Protocol 7*. To provide a concentrated phage stock for this procedure, make a plate lysate from the purified phage by inoculating at a high density such that total lysis of the Y1090 lawn occurs overnight. Cool the plate and add 4–5 ml of cold SM buffer. Gently shake or tip at 4°C for several hours. Recover the supernatant from the plate and transfer to a sterile tube with a drop of chloroform. This can be stored short term at 4°C, but for longer term storage add glycerol to 50% and maintain at −20°C. The plate lysate will contain 5×10^{10} to 10^{11} plaque-forming units (p.f.u.)/ml, but should be titred ready for lysogen generation.

Protocol 7. Generating a lysogen from purified phage stock

1. Inoculate one colony of Y1089[a] from a freshly streaked L-Broth/ampicillin plate into 5 ml of L-Broth plus maltose. Shake to an OD_{600} of 0.4 at 32°C.

2. Take 1 ml of culture and dilute 10-fold with L-Broth. 200 μl of this dilution will contain roughly 6.4×10^6 bacterial cells.

3. Infect 200 μl of cells at a multiplicity of 100–200 (i.e. 10 μl of a plate lysate with a titre of 10^{11} p.f.u./ml) for 20 min at room temperature.

4. Dilute each sample to 10 ml with L-Broth. Take 15 μl of this dilution and dilute again with 10 ml of L-Broth. This final dilution will have approximately 10^3 bacterial cells/ml. 200–300 μl can be spread on an L-Broth/ampicillin plate and grown overnight at 32°C.

5. Plates should contain 200–300 colonies. Replica plate 20–30 of these colonies on to two L-Broth/ampicillin plates in an asymmetric grid pattern. Grow one plate at 32°C (the master) and the other at 42°C overnight.

6. Colonies which grow at 32°C, but not at 42°C, can be assumed to contain an integrated copy of the purified phage.

[a] We have found that it is preferable to always maintain cultures of Y1089 at 32°C, except when inducing lysogenesis as in step 5.

Using this method, we have found that 10–50% of the replica-plated colonies in step 5 will be lysogens. The next step is to grow up a colony, induce protein synthesis, and prepare an extract. This is described in *Protocol 8*.

Protocol 8. Using a lysogen to make a crude protein extract

1. Inoculate an appropriate colony from the master plate into 5 ml of L-Broth plus ampicillin in a 50 ml Falcon tube. Shake vigorously at 32°C to an OD_{600} of approximately 0.5 (about 2 h).

2. Immerse the tubes in a 43°C water bath to raise the temperature rapidly, then transfer to a 43°C shaker for 20 min.

3. Add IPTG to 10 mM and shake for 1 h at 38°C.

4. Transfer 1.25 ml of culture into each of two 1.5 ml microcentrifuge tubes. Pellet the cells and remove the supernatant. Resuspend each pellet in 100 μl of buffer A (50 mM Tris (pH 7.5), 1 mM EDTA, 1 mM PMSF, 5 mM DTT; reference 6).

5. Recombine the two aliquots and lyse the cells with three rounds of freeze–thaw lysis. Add lysozyme to 0.5 mg/ml; incubate for 15 min on ice.

6. Add 55 μl of 5 M NaCl (to 1 M) and place on a rotary shaker for 15 min at 4°C. Spin for 30 min in a microcentrifuge at 4°C. The supernatant is a crude bacterial protein extract and will contain the cloned recombinant protein. Process wild type Y1089 alongside to provide a control extract.

The extract can be either diluted or dialysed versus buffer A supplemented with 10% glycerol to reduce the salt concentration for binding assays. The protein concentration of a dialysed sample is usually in the range of 3–5 µg/µl, but the exact value is readily determined using a commercial assay (Bio-Rad) and 1–5 µl should be sufficient for a binding assay.

Lambda lysogen extracts may be used to test the DNA-binding properties of a cloned factor by gel retardation or Southwestern analysis. It may be of value, for example, to compare the DNA-binding site preferences of the cloned factor with the known binding specificity of the factor in question. If antibodies are available, the identity of the cloned factor may also be checked immunologically using the λ lysogen extract and a gel retardation assay, since an antibody that recognizes the native protein will usually abolish or alter the mobility of the complex formed between the factor and its binding-site oligo-nucleotide. Alternatively, or in addition to the use of λ lysogen extracts, cloned factors may be translated *in vitro* following *in vitro* transcription of the cDNA, and the translated protein may be used for DNA-binding studies as detailed in Chapter 7. Obviously, the cloned factor will also need to be sequenced to check that there is an open reading frame of a suitable length to encode the factor of interest. Depending on the result, more cloning of 5' and 3' ends may be necessary. However, initial clones obtained from binding-site screens should contain the DNA-binding domain, so the derived amino acid sequence can be checked for known DNA-binding structures. This is particularly relevant if initial characterization of the factor and its binding site has indicated that this protein should be a member of one of the known families of DNA-binding proteins.

If a transcription factor is known to activate (or repress) transcription of a given gene, then the ability of the cloned factor to duplicate this effect may be tested by transient expression in eukaryotic cells in culture. In such experiments, a suitable reporter plasmid would contain a marker gene, such as chloramphenicol acetyl transferase (CAT) or luciferase, driven by a promoter containing binding-sites for the factor of interest. The reporter is co-transfected into cells with a second plasmid which expresses the cloned factor. The transcriptional activating/repressing activity of the cloned factor is then extrapolated from the activity of the reporter gene. Methods of analysing cloned transcription factors are described in more detail in Chapter 7.

A more time-consuming, but ultimately definitive, approach to decide if the cloned factor equates with the protein initially identified in nuclear extracts, is to raise antibodies against the newly cloned factor. This can be done by synthesizing suitable peptides (C- or N-terminal ones work best) determined from the derived amino acid sequence. Alternatively, all or part of the cloned factor may be expressed as a bacterial fusion protein (to facilitate purification, see Chapter 8) which may be used as an immunogen. Clearly, if these antibodies interact both with the cloned factor and with purified or partially purified preparations of the factor of interest, one is virtually home and dry!

References

1. Huynh, T. V., Young, R. A., and Davis, R. W. (1988). In *DNA Cloning: A Practical Approach*, vol. 1 (ed. D. M. Glover), pp. 49–78. IRL Press, Oxford.
2. Maekawa, T., Sakura, H., Kanei-Ishii, C., Sudo, T., Yoshimura, T., Fujisawa, J.-I., Yoshida, M., and Ishii, S. (1989). *EMBO J.*, **8**, 2023–2028.
3. Katagiri, F., Lam, E., and Chua, N.-H. (1989). *Nature*, **340**, 727–730.
4. Hai, T. W., Liu, F., Coukos, W. J., and Green, M. R. (1989). *Genes Dev.*, **3**, 2083–2090.
5. Poli, V., Mancini, F. P., and Cortese, R. (1990). *Cell*, **63**, 643–653.
6. Singh, H., LeBowitz, J. H., Baldwin, A. S., and Sharp, P. A. (1988). *Cell*, **52**, 415–423.
7. Kageyama, R. and Pastan, I. (1989). *Cell*, **59**, 815–825.
8. Klemsz, M. J., McKercher, S. R., Celada, A., Van Beveren, C., and Maki, R. A. (1990). *Cell*, **61**, 113–124.
9. Xiao, J. H., Davidson, I., Matthes, H., Garnier, J.-M., and Chambon, P. (1991). *Cell*, **65**, 551–568.
10. Williams, T. M., Moolten, D., Burlein, J., Romano, J., Bhaerman, R., Godillot, A., Mellon, M., Rauscher, F. J., and Kant, J. A. (1991). *Science*, **254**, 1791–1794.
11. Lum, L. S. Y., Sultzman, L. A., Kaufman, R. J., Linzer, D. I. H., and Wu, B. J. (1990). *Mol. Cell. Biol.*, **10**, 6709–6717.
12. Hurst, H. C., Masson, N., Jones, N. C., and Lee, K. A. W. (1990). *Mol. Cell. Biol.*, **10**, 6192–6203.
13. MacGregor, P. F., Abate, C., and Curran, T. (1990). *Oncogene*, **5**, 451–458.
14. Tsai, S. F., Martin, D. I. K., Zon, L. I., D'Andrea, A. D., Wong, G. G., and Orkin, S. H. (1989). *Nature*, **339**, 446–451.
15. Becker, D. M., Fikes, J. D., and Guarente, L. (1991). *Proc. Natl. Acad. Sci. USA*, **88**, 1968–1972.
16. Zhou, P. B. and Thiele, D. J. (1991). *Proc. Natl. Acad. Sci. USA*, **88**, 6112–6116.

6

Cloning transcription factors by homology

ALAN ASHWORTH

1. Introduction

Knowledge of the primary structure of different classes of transcription factors has revealed the modular nature of these proteins. Functional domains, such as DNA-binding motifs, are often highly conserved between different transcription factors. The sequence conservation of these domains can be taken advantage of to isolate clones related to a particular transcription factor. In some cases, these gene families may have more than 500 members (1).

2. Methods for cloning related transcription factors

Two methods are commonly used to isolate clones for related transcription factors. The traditional approach has been to use low-stringency hybridization with a probe for a member of the gene family. More recently, the polymerase chain reaction (PCR) using degenerate oligonucleotides has become popular. Both of these methods have advantages and disadvantages, and the choice of method would depend on the particular gene family to be analysed. Some examples of the different approaches are listed in *Table 2*.

2.1 Low-stringency hybridization

Low-stringency hybridization has been used to isolate clones for members of several transcription factor gene families (*Figure 1*). It is also particularly suitable for the isolation of the equivalent (orthologous) gene from phylogenetically diverse organisms. An obvious requirement for the use of this method is the availability of a cDNA or genomic clone for the factor in question. This can be generated by the PCR methods described in *Protocols 2* and *8*. A typical screening procedure is shown in *Protocol 1*. Note that the most important parameter is the signal-to-noise ratio and, particular attention,

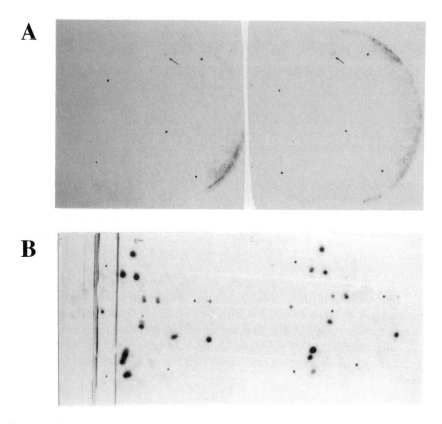

Figure 1. Low-stringency hybridization screening of a cDNA library. (A) Approximately 50 000 p.f.u. of an 8.5-day mouse embryo cDNA library in λgt10 were plated on to a 140 mm agar plate. Duplicate lifts were hybridized to a *Drosophila Kruppel* probe at low stringency (see *Protocol 1*). Filters were washed at low stringency and autoradiographed. Sharp dots are orientation marks. The single 'positive' is indicated by arrows. (B) The area corresponding to the positive plaque in (A) was excised and the phage eluted. Approximately 100 plaques were plated on a 90 mm dish and duplicate lifts hybridized as in (A).

therefore, should be paid to reducing background hybridization. Both genomic and cDNA libraries can be screened at low stringency, although the lower complexity of the latter leads to the isolation of fewer hybridization artefacts. Furthermore, eukaryotic genomes contain large numbers of pseudogenes and these can complicate the analysis of the products of a low-stringency screen.

2.1.1 Determining optimal conditions of hybridization

It is often useful to determine the optimal conditions for hybridization, before library screening is initiated. This is achieved by probing Southern and/or Northern blots under various stringency conditions. It is convenient to start with the lowest stringency conditions of hybridization and post-hybridization washing, expose the blot to film and then wash at higher stringency until the background is acceptable. Suggested conditions are given in *Protocol 1*. In addition, the screening of Northern blots of RNA isolated from different tissues can give an indication of which cDNA libraries could profitably be screened.

2.1.2 Choice of probe

Both cloned probes and labelled oligonucleotides have been used successfully. Cloned probes are generally used when the region of conservation is extensive. Probes from lower organisms such as *Drosophila* have frequently been used for the isolation of related mammalian genes containing motifs such as homeoboxes and zinc fingers (2, 3). Oligonucleotides are used when the region of homology is more restricted. As with PCR, degeneracy of the oligonucleotides can be reduced by considerations of codon preference (see Section 2.2.1 (c)) (4). Oligonucleotides are particularly useful for the isolation of clones containing multiple copies of a motif, as with the zinc finger (5) (*Table 2*).

Protocol 1. Low-stringency hybridization screening of cDNA and genomic libraries

1. Plate out a bacteriophage, or a plasmid or cosmid library, on to appropriate agar plates and take lifts with nylon membranes using standard protocols (6). Duplicate lifts are strongly recommended to distinguish genuine signals (see *Figure 1*).

2. Fix DNA to the filters according to the manufacturer's instructions.

3. Pre-wash the filters for 2 h in 1 litre of a solution containing 5 × SSC,[a] 0.5% sodium dodecyl sulphate, and 1 mM EDTA at 65°C. This removes much of the bacterial debris and considerably reduces the non-specific background. Briefly rinse the filters in the same solution.

4. Pre-hybridize the filters in either 6 × SSC,[a] 0.1% sodium dodecyl sulphate, 5 × Denhardt's reagent,[b] and 100 μg/ml salmon sperm DNA[c] at 50–65°C or the same solution containing 50% formamide at 30–42°C. At least 4 h of pre-hybridization are required for minimal background hybridization.

5. Add the ^{32}P-labelled probe and hybridize overnight at the above temperatures.

6. The filters should be washed extensively with large volumes and several changes of buffer. A reasonable starting point for washing is 2 × SSC[a]/ 0.1% sodium dodecyl sulphate at room temperature, although higher salt concentrations (and hence lower stringency) can be used. Care should be taken not to allow the filters to dry out before autoradiography as this makes subsequent higher stringency washing ineffective.

[a] 1 × SSC is 0.15 M NaCl and 0.015 M sodium citrate (pH 7.0).
[b] 50 × Denhardt's reagent is 5% Ficoll, 5% polyvinylpyrrolidone, and 5% bovine serum albumin (fraction V).
[c] Salmon sperm DNA should be sonicated and denatured before use (6).

2.2 Polymerase chain reaction approaches

The advent of PCR technology (7, 8) has facilitated the isolation of clones of many members of gene families which could not have been isolated by low-stringency hybridization. This is due, in part, to the minimal overall sequence homology required to construct PCR primers (*Figure 2*). These approaches also have the advantage that many samples can be screened for homology and that the procedure is relatively rapid. The disadvantage is the considerable number of artefactual sequences that can be generated.

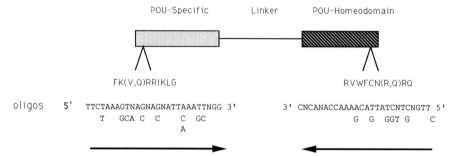

Figure 2. Degenerate PCR primers for POU-domain genes. The figure shows a schematic diagram of the structure of the bipartite POU domain. Degenerate primers are shown which are complementary to a consensus of previously isolated POU-domain proteins (19, 22).

2.2.1 Selection of PCR primers

The selection of PCR primers is of critical importance. Some general principles are:

(a) Select highly conserved sequences as primers. If possible, select them on the basis of conservation across species (the more distantly related the better) as well as conservation within the gene family.

(b) Choose sequences of minimal degeneracy (see *Table 1*):
  (i) peptides containing tryptophan and methionine are preferred as they are specific by unique codons
  (ii) avoid peptides containing leucine, arginine, and serine, if possible, as they result in a large increase in the number of degeneracies required.

(c) If the primer sequences are extremely degenerate, the degeneracy can reasonably safely be reduced by taking account of the under-representation of CpG dinucleotides in vertebrate genomes. For example, in a primer coding for a peptide containing histidine followed by glutamic acid, CA(T,C)GA(A,G) would normally be used. However, it is statistically unlikely that the CAC codon is used as it is followed by a G. Degeneracy at other positions could be reduced by utilizing codon

128

Table 1. Degeneracy of amino acid codons for selection of PCR primers

	Number of codons			
One	**Two**	**Three**	**Four**	**Six**[a]
Methionine (ATG)	Asparagine (AAQ)	Isoleucine (ATR)	Alanine (GCX)	Arginine (SGX)
Tryptophan (TGG)	Aspartic acid (GAQ)		Glycine (GGX)	Leucine (QTX)
	Cysteine (TGQ)		Proline (CCX)	Serine (VWX)
	Glutamic acid (GAP)		Threonine (ACX)	
	Glutamine (CAP)		Valine (GTX)	
	Histidine (CAQ)			
	Lysine (AAP)			
	Phenylalanine (TTQ)			
	Tyrosine (TAQ)			

Q = A or T; P = C or T; R = T, C, or A; S = C or A; V = A or T; W = G or C; X = T, C, G, or A.
[a] These amino acids are encoded by six codons but introduce eight (arginine and leucine) or sixteen (serine) degeneracies.

preference tables (4) or by including inosine at positions of degeneracy. During the course of the PCR, many mismatched primers are, in fact, incorporated into the product (see *Figure 3*).

(d) Residues at the 3′ end of the primers are most important in conferring specificity on the PCR. It is best, therefore, to have as many unambiguous residues at the 3′ end as possible. Alternatively, and perhaps surprisingly, it appears that T as the 3′ residue is relatively neutral when mismatched to G, C, or T (9).

(e) The length of the PCR product should optimally be 200–1000 bp, depending on the degeneracy of the primers. Fewer artifactual bands appear to be generated when amplifying small products with highly degenerate primers.

(f) If more than two regions in a protein are highly conserved, several sets of PCR primers can be constructed. This allows the use of 'nested' PCR which increases the specificity considerably. Multiple PCR primers have been used to clone a human TFIID gene, with the yeast sequence being used before it was known which parts of the protein are highly conserved (10).

Examples of the use of highly degenerate primers to isolate clones related to various transcription factors are discussed in Section 4.

2.2.2 PCR conditions

In general, it is worth trying the standard conditions given in *Protocol 2* first. These have frequently been successful for cross-species PCR (see reference 11). However, it is often necessary to optimize the efficiency of the PCR when highly degenerate primers are used. Some of the parameters that can be varied are given below.

i. Annealing temperature

In general, the lower the annealing temperature the more chance of non-specific amplification. This will, of course, depend on the degeneracy of the primers. Annealing temperatures as low as 37°C can be used with some highly degenerate primers with little artefactual amplification in a standard PCR (P. Denny and A. Ashworth, manuscript in preparation).

A method, called 'Touchdown PCR' (12) has been developed to attempt to reduce spurious amplification. In this procedure, amplification starts off above rather than at or below the expected annealing temperature. The annealing temperature is then decreased (by 1–2°C) every 1–2 cycles to a 'Touchdown' temperature at which 10–20 cycles are performed. This gives a selective advantage early in the PCR to highly matched sequences. This method was utilized in the cloning of the small subunit of the general transcription factor TFIIE (13). Highly degenerate PCR primers were designed based on the sequences of purified TFIIE peptides. A PCR starting at an

DP 1

5'	TTC	AAA	GTN	AGN	AGN	ATT	AAA	TTN	GG
	T	G	CA	C	C	C	G	C	
						A			
Oct-2	TTC	AAG	CAA	CGC	CGC	ATC	AAG	CTG	GG
	TTT	AAG	CTA	AGG	CGC	ATT	AAG	TTT	GG
	TTT	AAG	CTA	AGG	CGC	ATT	AAG	TTT	GG
Oct-1	TTC	AAA	CAA	AGA	CGA	ATC	AAA	CTT	GG
	TTT	AAG	CTA	AGG	CGC	ATC	AAA	CTT	GG
	TTC	AAG	CTA	CGG	CGC	ATC	AAG	CTT	GG
	aTT	tcG	aTT	AGG	AGC	ATC	AAG	CTT	GG
	TTT	AAG	CTA	AGG	CGC	ATC	AAG	TTT	GG
Oct-11	TTC	AAG	CAG	AGA	CGC	ATT	AAG	CTA	GG
	TTT	AAG	TAC	GGC	GGA	ATC	AAG	CTT	GG
	TTT	AAG	CTA	AGG	CGC	ATC	AAG	TTT	GG

DP 2

3'	GNG	TNT	GGT	TTT	GTA	ATA	ANA	GNC	AG
				C	C	CC	G C		A
Oct-2	GCG	TCT	GGT	TCT	GCA	ACC	GGC	GCC	AG
	GAG	TGT	GGT	TCT	GCA	ACC	G..	...	..
	GAG	TGT	GGT	TCT	GCA	ACC	AC.	...	..
Oct-1	GTG	TTT	GGT	TTT	GTA	ACC	GCC	GCC	AG
	GAG	TGT	GGT	TCT	GCA	ATC	GCC	GTC	AA
	GAG	TGT	GGT	TTT	GCA	ACC	GCC	GTC	A.
	GAG	TGT	GGT	TCT	GCA	ATC	GCC	GTC	A.
Oct-11	GAG	TCT	GGT	TCT	GCA	ATC	GAC	GCC	AA
	GAG	TGT	GGT	TCT	GTA	ACC	G..	...	..
	GAG	TGT	GGT	TCT	GTA	ACC	G..	...	..

Figure 3. Incorporation of mismatched primers into PCR products. Primer sequences incorporated into PCR products are compared with the relevant template sequences for the genes *Oct-1*, *Oct-2*, and *Oct-11*. DP1 and DP2 correspond to the forward and back degenerate primers in *Figure 2* (19, 22). Clearly, many of the PCR products have incorporated mismatched primers (mismatches are indicated by underlining). Some of the products have incorporated bases not present in the degenerate oligonucleotides (indicated by lower-case letters). Dots indicate bases not determined (A. S. Goldsborough, personal communication).

annealing temperature of 54°C and dropping by 2°C each cycle to a touchdown of 34°C was used. This annealing temperature was used for a further 40 cycles. The PCR product was sequenced and shown to be specific for TFIIE.

ii. Number of cycles of amplification

The use of an excessive number of cycles can lead to artefactual amplification. The minimum number necessary (usually 30–50) should be determined by removing an aliquot and testing the specificity of amplification every 5 cycles or so.

iii. Ramp time

The rate of cooling from the denaturing to the annealing stage can be varied on some PCR machines. In general, a slower ramp time should favour mismatch hybridization.

iv. PCR buffer composition

The major component of the PCR buffer that can usefully be varied is the concentration of Mg^{2+} ions. Multiple PCRs should be set up with final Mg^{2+} concentrations varying in 0.25 mM increments between 1.5 mM and 3.0 mM. This can have a considerable effect on the final yield of PCR product (see *Figure 4*) as well as the specificity, depending on the primers used.

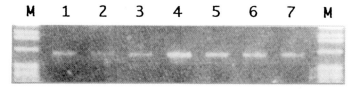

Figure 4. Determination of optimal Mg^{2+} concentration for PCR. PCRs were set up as described in *Protocol 2* using buffers differing only in their Mg^{2+} concentration. The Mg^{2+} concentrations were: 1, 1.5 mM; 2, 1.75 mM; 3, 2.0 mM; 4, 2.25 mM; 5, 2.5 mM; 6, 2.75 mM; 7, 3.0 mM. The optimal concentration for this set of primers was 2.25 mM.

2.2.3 Choice of template

Genomic DNA, cDNA reverse-transcribed from RNA, and cDNA libraries are all suitable for PCR with degenerate primers.

i. Genomic DNA

About 0.25–1.0 µg genomic DNA can be included in a standard PCR reaction (*Protocol 2*). Clearly, large introns should not be present within the region to be amplified if this approach is to be successful. In addition, there is the danger that pseudogenes, of which there are many in vertebrate genomes, will also be amplified when genomic DNA is used as template. The advantage is that one does not need prior knowledge of where the homologous gene is expressed.

ii. cDNA

This can be synthesized using the method described in *Protocol 8*. The amount of cDNA included in the PCR should be varied as the yield of reverse transcription reactions is variable. Priming of the cDNA can be with oligo dT, random hexamers, or a specific oligonucleotide. In addition, the use of the RACE (rapid amplification of cDNA ends) protocol should allow PCR with only a single degenerate oligonucleotide, although we have not tried this.

iii. cDNA libraries

High-titre bacteriophage and plasmid cDNA libraries can be used as templates for PCR. This has several advantages. The PCR product generated can be used directly to screen the library for longer clones by filter hybridization.

The fact that the insert in the cDNA library is flanked by known vector sequences allows alternative PCR approaches to be used. PCR can be performed with one degenerate primer and a vector primer ('in-and-out' PCR). If a specific product is generated, it can be sequenced directly using the vector primer (see below). An extension of this technique is to fractionate a cDNA library and analyse the fractions by PCR, either with two degenerate primers or by the in-and-out procedure (*Protocol 3*). The PCR product itself can then be used as a hybridization probe to screen the appropriate fraction.

Protocol 2. Polymerase chain reaction—standard conditions

1. Assemble the components of the reaction at room temperature as follows:
   - 10 × PCR buffer (150 mM Tris–HCl (pH 8.8), 600 mM KCl, 22.5 mM $MgCl_2$)   5 μl
   - 10 mM dNTPs (Pharmacia or Boehringer Mannheim)   1.25 μl
   - of each primer at 10 pmol/μl   1 μl
   - Taq Polymerase (Perkin-Elmer Cetus, many other sources are also suitable)   2.5 units
   - H_2O to 49 μl

   It is often convenient to prepare a master mix of these components which can then be divided into aliquot parts before addition of template.

2. Add the template last to reduce chances of cross-contamination. Thorough mixing appears to be unnecessary. The template may be genomic DNA (50–500 ng), cDNA (see *Protocol 8*), or cDNA libraries (1–5 μl). Care should be taken over the buffer in which the template is added as the addition of extraneous Mg^{2+} ions and EDTA could alter the efficiency of the PCR.

3. Overlay the reactions with two drops (50–100 μl) of light mineral oil (Sigma No. M-3516) and close the tubes.

Protocol 2. *Continued*

4. Subject the samples to cycles of denaturation, annealing, and extension on a thermal cycler. Exact conditions will depend on the cycler used, but the following conditions have proved suitable for a range of primers and templates. Thirty cycles of:

   (a) denaturation at 94°C for 30 sec

   (b) annealing at 50–55°C for 30 sec

   (c) extension for 60 sec at 72°C

   followed by 5 min at 72°C. The extension time has been found to be sufficient for the synthesis of greater than 1.5 kb.

5. The PCR products can be loaded, with care, directly on to agarose (1–1.6% regular agarose or 3% NuSieve GTG agarose (FMC) + 1% regular agarose) gels or polyacrylamide (10–15%) gels. If the mineral oil is a problem, it can be extracted by the addition of 100 μl chloroform. Alternatively, place the PCR tubes at −20°C until the aqueous phase freezes. The mineral oil remains liquid and can be removed with a pipette. This is particularly useful if the PCRs are performed in microtitre well dishes (14).

Protocol 3. Screening libraries by PCR

1. Plate out cDNA or genomic library in a λ vector at a density of 5000–40000 plaques per 140 mm bacterial dish.

2. Grow for 12–16 h to near confluence. Lifts can be taken at this stage with nylon filters, or one can proceed directly to the next step.

3. Pipette approximately 8 ml of 100 mM NaCl, 10 mM MgCl$_2$, 10 m Tris–HCl (pH 8.0) on to the surface of each of the plates and incubate at room temperature with shaking for 1–2 h.

4. Recover the buffer containing the amplified fractions of the library and analyse by PCR as described in *Protocol 2*.

5. PCR products are subcloned and sequenced. In addition, they can be labelled with ^{32}P and used as probes to screen the filters generated in Step 2. Alternatively, the positive fractions of the library can be replated and fresh filters generated for hybridization with the ^{32}P-labelled probe.

2.2.4 Sequencing PCR products

Most PCR products derived by the methods described above will be mixtures of products of more than one gene. As such they cannot be sequenced directly and must be subcloned.

i. Subcloning PCR products

Various methods are available for subcloning PCR products. The addition of restriction enzyme recognition sites to the end of the oligonucleotide primers used for amplification should allow the cloning of the digested PCR products into similarly digested vectors. However, many restriction enzymes fail to cleave efficiently if their recognition sequences are located too close to the ends of the oligonucleotide. We find, however, that direct cloning of PCR products either by blunt-end ligation (*Protocol 4*) or by ligation into T-tailed vectors (15) (*Protocol 5*) is very efficient. Both these approaches can be used for the subcloning of essentially any PCR product. T vectors take advantage of the A residue frequently added to the 3' end of the PCR product during the PCR itself.

Protocol 4. Cloning PCR products by blunt-end ligation

1. PCR products electrophoresed in agarose or polyacrylamide gels are purified using the Geneclean or Mermaid (BIO101) procedures. Mermaid is used for fragments of less than 400 bp. The DNA is eluted into 13 μl H_2O.

2. Add 2 μl 10 × kinase buffer (700 mM Tris–HCl (pH 7.5)/100 mM $MgCl_2$), 2 μl 100 mM dithiothreitol (DTT), 2 μl 10 mM ATP, and 1 μl (~5 units) T4 polynucleotide kinase. Mix and incubate at 37°C for 30 min.

3. Add 3 μl 0.5 mM dNTPs and 1 μl Klenow polymerase (5 units). Mix and incubate at room temperature for 15 min.

4. Purify DNA by the Geneclean or Mermaid procedures. Elute into 16 μl H_2O.

5. Mix with 2 μl 10 × ligase buffer (Boehringer), 1 μl (5–10 units) T4 DNA ligase and 1 μl (20 ng) of a suitable blunt-ended vector that has been treated with alkaline phosphatase. A control ligation containing H_2O instead of the PCR product should also be set up to assess the level of vector background. A suitable vector is pBluescript (Stratagene) cut with *Eco*RV and treated with calf intestinal alkaline phosphatase.

6. After a minimum of 2 h, transform 10 μl of the ligations into *E.coli* competent cells (JM83 or XL-1Blue, or others for pBluescript) and plate on to L-agar plates containing appropriate supplements such as ampicillin and X-gal.

Protocol 5. Cloning PCR products using T-tailed vectors

1. Digest 2 μg pBluescript or other suitable plasmid with *Eco*RV and purify the linearized plasmid by agarose gel electrophoresis and the Geneclean procedure.

Protocol 5. *Continued*

2. Incubate the digested plasmid with Taq polymerase (2 units) in 1 × PCR buffer (see *Protocol 2*) in the presence of 2 mM TTP for 2 h at 72°C. Purify by the Geneclean procedure and elute into 50 µl H$_2$O.

3. Set up ligations of 1 µl of this vector with aliquots of the gel-purified PCR products and transform into suitable bacteria as in steps 5 and 6 of *Protocol 4*.

---

ii. Sequencing subcloned PCR products

It is most convenient to sequence directly from bacterial colonies harbouring plasmids containing the subcloned PCR product. This is done by PCR on the bacterial colonies with primers flanking the cloning site in the plasmid used (*Protocol 6*). The PCR product may then be sequenced directly using one of the methods (such as *Protocol 7*) available for this. These procedures allow the rapid sequence analysis of the products of the PCR (*Figure 5*).

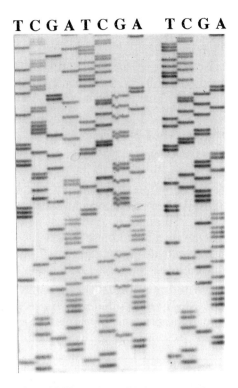

T C G A T C G A   T C G A

Figure 5. Direct sequencing of PCR products. DNA was amplified as described in *Protocol 2* and sequenced directly as described in *Protocol 7*. The sequencing reactions were electrophoresed on a thin polyacrylamide/urea wedge gel and the gel was dried and autoradiographed.

Protocol 6. Amplification of DNA from bacterial colonies—'wiggles'

1. Make up PCR mixtures as in *Protocol 2*. Primers should be complementary to regions flanking the restriction site used for subcloning. They should be further than 20 bp away from the cloning site to enable the full sequence of the insert to be determined (*Protocol 7*).

2. Using a disposable microbiological loop, pick a bacterial colony, twirl it in the PCR mix for 2–3 sec and then place the loop into a culture tube containing 5 ml L-Broth + 50 μg/ml ampicillin.

3. Add mineral oil to the PCR tubes and cycle as in *Protocol 2*. The cultures are shaken at 37°C for 5 h to overnight.[a]

4. The PCRs are analysed by agarose gel electrophoresis. Bands are excised, the DNA purified by the Geneclean or Mermaid (BIO101) procedures and sequenced directly by the method described in *Protocol 7*.

---

[a] Alternatively, PCR can be performed on the bacterial culture after a few hours of growth as follows: centrifuge 100 μl bacterial culture for 1 min in a microfuge, discard the supernatant and resuspend the pellet in 50 μl H_2O. Boil the tube for 5 min, centrifuge for 1 min, and use 1 μl of the supernatant in the PCR.

Protocol 7. Direct sequencing of PCR products

1. Mix 7 μl (0.25–1.0 μg) PCR product (*Protocol 6*) with 1 μl (10 pmol) of the appropriate sequencing primer and heat at 95–100°C for 5 min. This is most conveniently done in a heating block. One of the primers from the PCR (*Protocol 6*) or an internal primer may be used for sequencing.

2. Place the tube on dry ice for 1 min to freeze, then centrifuge briefly (2–3 sec) in a microcentrifuge.

3. Add 7.5 μl of a previously prepared reaction mix composed of[a]:

   - 5 × sequencing buffer (200 mM Tris–HCl (pH 7.5),
     100 mM $MgCl_2$, 250 mM NaCl)   2 μl
   - 100 mM DTT   1 μl
   - diluted labelling mix (7.5 μM dCTP, dGTP, and TTP;
     this is diluted 1:4 in water)   2 μl
   - [α-^{35}S]dATP (Amersham, 1000 Ci/mmol)   0.5 μl
   - T7 DNA polymerase (Pharmacia; diluted in 10 mM Tris–HCl
     (pH 7.5), 5 mM DTT, 0.5 mg/ml bovine serum albumin)   2 μl (2 units)

4. Centrifuge briefly and incubate at room temperature for 2 min. During this time add 3.5 μl of the mixture to the lip of each of four microcentrifuge tubes containing T, C, G, or A termination mixes.[b] The termination

Protocol 7. *Continued*

reaction is initiated by brief centrifugation. The tubes are placed in a 37°C water bath and incubated for 5 min.

5. Following addition of 4 μl of formamide/dye mix (95% formamide, 20 mM EDTA, 0.05% bromophenol blue, 0.05% xylene cyanol FF) the reactions are heated at 95°C for 2–3 min and loaded on to a 6% polyacrylamide/7 M urea sequencing gel.

---

[a] 0.5 μl MnCl₂/sodium isocitrate buffer (16) can be included if sequences close to the primer site are required.

b Termination mixes have the following composition:
- T termination mix: 80 μM dNTPs, 50 mM NaCl, and 8 μM ddTTP
- C termination mix: 80 μM dNTPs, 50 mM NaCl, and 8 μM ddCTP
- G termination mix: 80 μM dNTPs, 50 mM NaCl, and 8 μM ddGTP
- A termination mix: 80 μM dNTPs, 50 mM NaCl, and 8 μM ddATP.

2.2.5 Artefacts

Great care should be taken with PCR approaches to avoid and recognize artefacts.

(a) The PCR is very prone to artefacts due to its extreme sensitivity. Hence the usual precautions (see *PCR, A Practical Approach* in this series) should be taken to avoid contamination from plasmid stocks containing amplifiable sequences. If necessary, pipettes, tubes and buffers can be treated with UV light to eliminate contamination from nucleic acids (17).

(b) Depending on the conditions used, the use of degenerate primers can produce large numbers of artefactual sequences due to mispriming. These can be reduced by careful titration of the PCR conditions (see Section 2.2.2 above).

(c) The inherent error rate of Taq polymerase is significant and all sequences should be assembled from more than one PCR product. The use of thermostable DNA polymerases with much lower error rates (such as *Pyrococcus* DNA polymerase) should circumvent this problem.

(d) Often a specific or artefactual product will amplify preferentially, making the isolation of other related sequences difficult. This can be partially alleviated by restriction digestion of the PCR product with an enzyme specific for the preferentially amplified product.

(e) Occasionally, sequence analysis of a single subcloned PCR product can yield a mixed sequence where bands are observed in two tracks of a sequencing gel. This is almost certainly due to cloning of a heteroduplex. This would be replicated after transformation, resulting in the bacterium in question containing two distinct plasmids.

(f) Chimeric or recombinant sequences can be formed in the PCR if closely related sequences are present. This will be a particular problem at high product concentration towards the end of the PCR. Increasing the elongation time has been reported to reduce the frequency of this phenomenon.

(g) Yeast sequences are represented in some commercially available cDNA libraries, probably due to the use of yeast RNA as a carrier in cDNA preparation (18). As a consequence, care should be taken to confirm that the cDNA is derived from the appropriate organism.

(h) The isolation of a PCR product from a particular cDNA sample does not necessarily indicate that this gene is significantly expressed in that tissue. We isolated a novel octamer binding protein gene, *Oct-11*, by PCR from testis cDNA (19). Subsequently, however, we were unable to detect expression of *Oct-11* in the testis by Northern blotting. Furthermore, we were unable to isolate *Oct-11* cDNA clones after extensive screening of testis cDNA libraries (A. S. Goldsborough and A. Ashworth, in preparation). We presume that the extreme sensitivity of the PCR technique led to the isolation of the *Oct-11* gene which is not significantly expressed in the testis. Thus care should be taken to confirm that the isolated PCR product is meaningfully expressed. Specific PCR primers can be designed and the expression of the gene studied by RT-PCR (*Protocol 8*) and the results can be confirmed by Northern blotting.

Protocol 8. Reverse transcription of RNA for PCR analysis

1. Prepare RNA (total or poly A enriched) from tissues or cells by any standard method.

2. Set up reverse transcription by mixing the following components on ice:

- 5 × reverse transcription buffer (Gibco-BRL)   4 μl
- 10 mM dNTPs   1 μl
- 100 mM DTT   2 μl
- primer–oligo dT_{12-18}, random hexamers, or a specific primer
                                                                          1 μl (10 pmol)
- human placental ribonuclease inhibitor   0.5 μl (5 units)
- H_2O 10 μl
- Moloney murine leukaemia virus reverse transcriptase (Gibco-BRL)
                                                                        0.5 μl (100 units)

3. Heat 1 μl RNA at 70°C for 2 min, chill on ice and add to the reaction mix. Incubate at 37°C for 30 min.

4. Dilute to 50–100 μl with H_2O. Include various amounts (usually 1 μl) of the cDNA in PCR reactions with suitable primers. Analyse the products by gel electrophoresis.

Table 2. Examples of novel transcription factors isolated by PCR or low-stringency hybridization

Class of transcription factor	Conserved domain	Notes	References
Homeobox proteins	60 amino acid DNA-binding domain (homeobox)	Conserved domain originally described in *Drosophila* homeotic genes. Present in a wide range of eukaryotic organisms. Low-stringency hybridization with *Drosophila* probes, hybridization with degenerate oligonucleotides, and PCR have been used to isolate members of this gene family	2, 20, 21
Zinc-finger proteins	28–30 amino acid motif frequently tandemly repeated and based on the motif $CX_2CX_3FX_5LX_2HX_3HTGEKP(F,Y)$	The zinc-finger motif was originally described in transcription factor IIIA from *Xenopus*. The *Drosophila* *Kruppel* gene and other probes have been used to isolate colonies for zinc-finger proteins. In addition, hybridization with degenerate oligonucleotides and PCR have also been used. Mammalian genomes may contain more than 500 genes of the zinc-finger class	1, 3
POU-domain proteins	Tripartite structure; 75–80 amino acid POU-specific domain, a non-conserved linker, and a 60 amino acid carboxy-terminal divergent homeodomain	Conserved region originally described in mammalian proteins Pit-1, Oct-1, and Oct-2 and in the product of the nematode *Unc-86* gene, hence referred to as POU family of genes. PCR has been utilized to isolate large numbers of POU genes. Low-stringency hybridization has also been used	19, 22–24
Helix–loop–helix proteins	Two segments of 12–15 amino acids capable of forming amphipathic a helices connected by a non-conserved loop region of varying length. Associated with a 10–20 amino acid basic region	Homology first identified among *c-myc*, the muscle determination gene *MyoD*, and genes of the *Drosophila* achaete–scute complex. Many other members now identified.	25, 26

3. Some examples of the cloning of transcription factors by homology

Some examples of the use of low-stringency hybridization and PCR to isolate members of transcription factor families are given in *Table 2*. This list is not intended to be exhaustive, but represents the best-documented examples of the use of these techniques.

Acknowledgements

Research in my laboratory is supported by the Cancer Research Campaign, the Medical Research Council and the Wellcome Trust. Thanks to Andy Goldsborough and Paul Denny for unpublished information and evolving many of the protocols described here.

References

1. Ashworth, A. and Denny, P. (1991). *Mammalian Genome*, **1**, 196.
2. McGinnis, W., Garber, R. L., Wirz, J., Kuroiwa, A., and Gehring, W. J. (1984). *Cell*, **37**, 403.
3. Chowdhury, K., Deutsch, U., and Gruss, P. (1987). *Cell*, **48**, 771.
4. Lathe, R. (1985). *J. Mol. Biol.*, **183**, 1.
5. Bellefroid, E. J., Lecocq, P. J., Benhida, A., Poncelet, D. A., Belayew, A., and Martial, J. A. (1989). *DNA*, **8**, 377.
6. Sambrook, J. Fritsch, E. F., and Marcatis, T. (ed.) (1989). *Molecular Cloning, a Laboratory Manual* (second edition). Cold Spring Harbor Press, Cold Spring Harbor, NY.
7. Saiki, R., Scharf, S., Faloona, F., Mullis, K. B., Horn, G. T., Erlich, H. A., and Arnheim, N. (1985). *Science*, **230**, 1350.
8. Saiki, R. K., Gelfand, D. H., Stofel, S., Scarf, S. J., Higuchi, R., Horn, G. T., Mullis, K. B., and Erlich, H. A. (1988). *Science*, **239**, 487.
9. Kwok, S., Kellogg, D. E., McKinney, N., Spasic, D., Goda, L., Levenson, C., and Sninsky, J. J. (1990). *Nucleic Acids Res.*, **18**, 999.
10. Hoffmann, A., Sinn, E., Yamamoto, T., Wang, J., Roy, A., Horikoshi, M., and Roeder, R. G. (1990). *Nature*, **346**, 387.
11. Ashworth, A., Rastan, S., Lovell-Badge, R., and Kay, G. (1991). *Nature*, **351**, 406.
12. Don, R. H., Cox, P. T., Wainwright, B. J., Baker, K., and Mattick, J. S. (1991). *Nucleic Acids Res.*, **19**, 4008.
13. Peterson, M. G., Inostroza, J., Maxon, M. E., Flores, O., Admon, A., Reinberg, D., and Tjian, R. (1991). *Nature*, **354**, 369.
14. Lennon, G. G., Drmanac, R., and Lehrach, H. (1991). *Biotechniques*, **11**, 185.
15. Marchuk, D., Drumm, M., Saulino, A., and Collins, F. S. (1991). *Nucleic Acids Res.*, **19**, 1154.
16. Tabor, S. and Richardson, C. C. (1989). *Proc. Natl. Acad. Sci. USA*, **86**, 4076.
17. Sarker, G. and Sommer, S. S. (1990). *Nature*, **343**, 27.

18. Lovett, M., Kere, J., and Hinton, L. M. (1991). *Proc. Natl. Acad. Sci. USA*, **88,** 9628.
19. Goldsborough, A. S., Ashworth, A., and Willison, K. R. (1990). *Nucleic Acids Res.*, **18,** 1634.
20. Crompton, M. R., MacGregor, A. D., and Goodwin, G. H. (1991). *Leukemia*, **5,** 357.
21. Singh, G., Kaur, S., Stock, J. L., Jenkins, N. A., Gilbert, D. J., Copeland, N. G., and Potter, S. S. (1991). *Proc. Natl. Acad. Sci. USA*, **88,** 10706.
22. He, X., Treacy, M. N., Simmons, D., Ingraham, H. A., Swanson, L. W., and Rosenfeld, M. G. (1989). *Nature*, **340,** 35.
23. Herr, W., Sturm, R. A., Clerc, R. G., Corcoran, L. M., Baltimore, D., Sharpe, P. A., Ingraham, H. A., Rosenfeld, M. G., Finney, M., Ruvkun, G., and Horvitz, H. R. (1988). *Genes Dev.*, **2,** 1513.
24. Scholer, H. R., Ruppert, S., Suzuki, N., Chowdhury, K., and Gruss, P. (199). *Nature*, **344,** 435.
25. Benezra, R., Davis, R. L., Lockshon, D., Turner, D. L., and Weintraub, H. (1990). *Cell*, **61,** 49.
26. Davis, R. L., Weintraub, H., and Lassar, A. B. (1987). *Cell*, **51,** 987.

7

Analysis of cloned factors

ROGER WHITE and MALCOLM PARKER

1. Introduction

The aim of this chapter is to describe methods used to investigate the functional properties of cloned transcription factors. The first section will be concerned with assay systems to measure the ability of the factor to stimulate transcription of an appropriate reporter gene using co-transfection into heterologous cells. A number of overexpression systems will then be discussed and the advantages and disadvantages of each for analysing transcription factor function will be evaluated. Finally, we will describe techniques used in the mapping and analysis of individual domains within the factor and the application of point mutagenesis in the identification of critical residues required for transcription factor function.

The majority of transcription factors contain functional domains for DNA binding and transcriptional activation, and many bind to DNA either in the form of homodimers or as heterodimers with other proteins. A number, namely the nuclear hormone receptors, depend on the binding of a hormonal ligand for their activity. The chapter will be illustrated by reference to studies using the mouse oestrogen receptor, a member of the nuclear hormone receptor family of transcription factors, and will compare the functional properties of the receptor with those of other transcription factor proteins. The nuclear hormone receptors are multifunctional proteins which have the ability to respond to signals from the extracellular environment and to act as transcription factors by directly regulating gene expression. These proteins, therefore, contain a number of functional properties, including ligand binding, nuclear localization, DNA binding, and transcriptional activation. The residues required for sequence-specific DNA binding have been identified by mutational analysis and structural studies and form a single domain which is able to function in isolation. In the oestrogen receptor, hormone binding and hormone-inducible transcriptional activation are localized in the same region of the protein and these sequences overlap in part with residues implicated in forming a major part of the dimer interface. In the identification of specific residues involved in individual functions it is important to ensure that mutational analysis does not disrupt the overall protein structure. The co-localization of functional properties in the hormone-binding domain of the oestrogen receptor allows a detailed analysis of the protein by mutagenesis since changes may be introduced to alter one function while the others are unaffected.

2. Identification of a cloned factor by co-transfection assays

A detailed functional analysis of a proposed transcription factor initially requires confirmation of both the identity of the factor and its ability to regulate gene expression. Identification may be accomplished directly by using a co-transfection assay, where the factor and an appropriate target gene are introduced into heterologous cells in which the factor is normally absent or expressed at very low levels. This type of analysis demands that the specific DNA sequence of the response element of the transcription factor has been determined. Identification of a cloned transcription factor may also be confirmed using *in vitro* DNA-binding assays by demonstrating with mutant binding sites that DNA-binding activity correlates with functional activity determined in transient transfection experiments. The methods described in the following section have been used in the analysis of the transcriptional activity of the mouse oestrogen receptor following transient transfection into a number of receptor negative cell lines. The major considerations to be taken into account in the design of this type of procedure are:

(a) the selection of a suitable cell line

(b) the choice of expression vector

(c) the type of target or reporter gene

(d) the method of transfection

2.1 Cell lines

There are two main criteria for selection of cell lines in which to study transcription factor activity:

(a) The cell line should either lack or have very low endogenous levels of the factor.

(b) The cells should be able to be transfected easily and efficiently.

Cell lines which have been used to study the mouse oestrogen receptor include mouse L-cells, NIH-3T3 cells, COS-1, and Hela. All of these cell lines lack endogenous oestrogen receptors, and suitable methods for the introduction of DNA by transfection have been established. Detailed protocols for transient transfection of NIH-3T3 cells and COS-1 cells will be described later (*Protocols 1* and 2).

A third consideration in the selection of suitable cell lines for studying an inducible transcription factor is the ability to maintain the cells in the absence of the inducer. For example, the analysis of hormone-dependent transcriptional activity by the oestrogen receptor requires that the cells are withdrawn from hormone prior to transfection. This includes the maintenance of cells in media without the pH indicator phenol red, which has been shown to be weakly oestrogenic (1) and the use of serum from which steroids have been removed by treatment with dextran-coated charcoal (DCC) (2). These treatments are not required when the cell line used can be maintained in serum-free conditions and defined media for the duration of the experiment.

2.2 Organization of expression vectors

The most important feature of an expression vector is a strong promoter which is active in a wide variety of eucaryotic cells. The vector will also contain splicing and termination sequences to generate stable and functional mRNA, a polylinker to allow insertion of a cDNA clone of the transcription factor, and an origin of replication and antibiotic resistance gene for growth and amplification in bacterial cells. A number of different expression vectors have been developed and are suitable for the expression of transcription factors, with the majority of vectors containing promoters and enhancers of viral origin (3).

Promoter elements contain DNA sequences which determine both the site and the rate at which transcription is initiated. Enhancer elements consist of binding sites for proteins involved in the stimulation of transcription. Many viral promoters and enhancers have a wide host range and, therefore, stimulate transcription efficiently in a variety of cell types. These include the SV4O early promoter, the Moloney murine leukemia virus long terminal repeat (LTR), and the human cytomegalovirus major immediate early promoter. Other enhancers may be cell-type specific or, as in the case of the mouse mammary tumour virus LTR, stimulate transcription efficiently only in the presence of an inducer. This type of expression system may be advantageous if expression of the protein is toxic to the transfected cells.

A cDNA clone containing the entire coding sequence of the mouse oestrogen receptor has been inserted as an *Eco*R1 fragment of 1.86 kb in size into the unique *Eco*R1 site in the expression vector pJ3Ω (4). The vector consists of the SV40 early promoter and enhancer elements, the viral origin of replication, and SV40 termination and polyadenylation sequences. The cDNA clone includes 11 bp of 5' non-coding sequence and 50 bp and 3' non-coding sequence. The mRNA transcribed from pJ3ΩMOR (*Figure 1*) is approximately 2.3 kb in size and contains a single open reading frame which is translated to give a polypeptide of 599 amino acids.

2.3 Reporter genes and control plasmids

A reporter gene contains three major regions: a binding site (or sites) forming a response element for the transcription factor being analysed, a basal

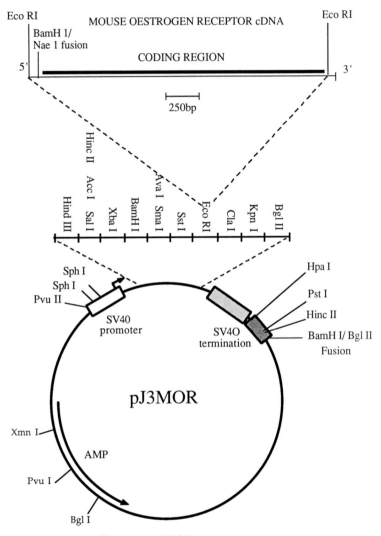

Figure 1. pJ3MOR expression vector.

promoter to provide binding sites for general transcription factors and to determine correct initiation of transcription, and a suitable marker gene for which a simple assay system is available. A number of different reporter genes have been used in co-transfection assays to study transcription factor activity. These include chloramphenicol acetyl transferase (CAT) (5), β-galactosidase (6), luciferase (7), and β-globin (8). Transcriptional activation by the mouse oestrogen receptor has been analysed using a reporter consisting of a single oestrogen response element (ERE) contained within a 32 bp

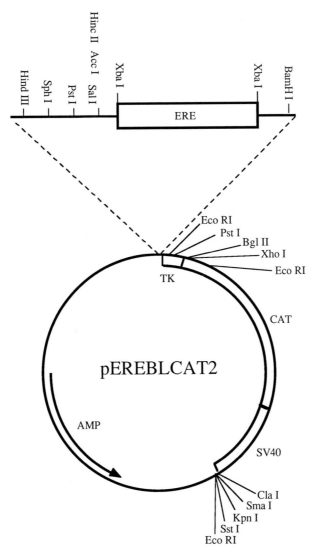

Figure 2. Reporter plasmid pEREBLCAT2.

oligonucleotide isolated from the vitellogenin A2 gene from *Xenopus laevis* (9). The ERE has been characterized in detail and the binding site consists of the 13-nucleotide sequence GGTCANNNTGACC which forms a palindromic response element. In the reporter plasmid pEREBLCAT$_2$ (*Figure 2*) the 32 bp oligonucleotide has been synthesized and cloned as an *Xba*1 fragment into the polylinker of the vector pBLCAT2 (10). This vector contains the basal promoter of the herpes simplex virus thymidine kinase (*tk*) gene from −105 to +51 linked to the coding region of the bacterial CAT gene and

147

is, therefore, suitable for use with any response element. Chimeric CAT fusion genes are particularly useful because there is no endogenous CAT activity in eucaryotic cells and the CAT enzyme can be monitored by a rapid and sensitive assay.

One potential criticism of assays which depend on the measurement of an enzyme activity such as CAT is that this is not a direct measurement of transcriptional activation. A number of studies have shown, however, that the level of CAT activity correlates with the steady-state level of mRNA produced. Levels of mRNA may be analysed directly using the techniques of primer extension or RNase mapping, and these methods also confirm correct initiation of transcription. This is more easily accomplished using β-globin as a reporter gene which produces relatively stable RNA transcripts compared with CAT. These assay systems, however, are relatively time consuming and contrast with enzyme assays which are both sensitive and simple to perform and may be used to analyse a large number of samples. Improvements in CAT vectors have also recently been introduced to minimize background transcription derived from cryptic promoters within vector sequences (11). Analysis of transcription factor function in a co-transfection experiment also requires the introduction of an internal control in addition to the expression vector and the reporter plasmid. This is particularly important where comparison is being made either between mutated forms of the transcription factor and the normal protein, or with a factor where transactivation is measured in the presence and absence of an inducer. In these experiments, the data obtained from the reporter must be normalized for both transfection efficiency and for general effects on transcription. A suitable control plasmid which has been used to normalize data in the analysis of oestrogen receptor activity consists of the firefly *Photinus pyralis* luciferase gene inserted into the eucaryotic expression vector pJ3Ω. Luciferase activity is measured using an assay in which light is generated from conversion of the substrate luciferin in the presence of ATP and Mg^{2+}. The reaction is carried out in a luminometer cuvette in a LKB-Wallac 1251 luminometer and the peak light emission recorded. A simple procedure is used to obtain cell extracts suitable for both CAT and luciferase analysis; a protocol for these assays is described at the end of the next section following protocols for methods of transfection.

2.4 Methods of transfection

A number of different methods have been developed for the introduction of DNA into cells, all of which are suitable for the analysis of transcription factor activity. The method chosen depends to a large extent on the particular cell line to be transfected. The most commonly used methods are:

(a) calcium phosphate-mediated transfection

(b) DEAE-dextran

(c) electroporation

(d) lipofection

Calcium phosphate-mediated transfection (12) may be used on a wide variety of cell lines, however, the efficiency of uptake of DNA may vary between cell types. Electroporation, which involves applying a brief high-voltage electric pulse to the cells to generate pores in the plasma membrane, results in a large percentage of cells taking up DNA, although the precise conditions for each cell type must be determined empirically (13). Lipofection (14) has been found to be the method of choice for a number of cell lines which transfect inefficiently by other techniques.

This section will describe two methods of transfection which have been used to analyse oestrogen receptor function in heterologous cells. The first method involves transfection of NIH-3T3 cells using calcium phosphate co-precipitation. This will be followed by a method describing the introduction of DNA into COS-1 cells by use of electroporation. In all cases, the DNA used should be purified from contaminating RNA and nicked or open circular DNA by use of caesium chloride gradients.

Protocol 1. Transient transfection using calcium phosphate co-precipitation

1. Prepare the following solutions:
   - $2 \times$ Hepes-buffered saline (HBS)[a]
   - 70 mM Na_2HPO_4
   - 70 mM NaH_2PO_4
   - 2 M calcium chloride (tissue-culture grade)

2. Plate cells 24 h before transfection in 6 cm diameter tissue-culture dishes at a density at which the cells will be approximately 80% confluent in 3–4 days (2×10^5 cells/dish for NIH-3T3 cells). If it is necessary to use larger dishes of cells, the number of cells plated and the volumes given should be adjusted accordingly.

3. The following day, remove the medium (including any dead cells) and replace with 4 ml of fresh medium for each 6 cm dish.

4. Prepare the following solutions to make 1 ml of precipitate which is sufficient for two 6 cm dishes:
   - Solution A: mix 5 μl 70 mM Na_2HPO_4 and 5 μl of NaH_2PO_4. Add 500 μl $2 \times$ HBS.
   - Solution B: prepare 20 μg DNA (including 10 μg of the reporter, 1 μg of expression vector, 2 μg internal control plasmid, and 7 μg pJ3Ω as carrier) in a total volume of 440 μl sterile distilled water. Add 60 μl 2 M $CaCl_2$.

Protocol 1. *Continued*

5. Add solution B to solution A at a constant rate of approximately one drop per second while continuously bubbling air through the solutions to aid mixing.

6. Leave the mixture for 30 min at room temperature to allow the precipitate to form.

7. Add 500 µl of the suspension to each dish of cells.

8. Place the dishes of cells in a humidified incubator at 37°C for 6 h. A fine-grained precipitate will form and attach to the surfaces of the cells.

9. Aspirate the medium from the cells and wash each dish of cells with 5 ml of medium lacking serum. This washing procedure is repeated until all traces of precipitate have been removed and normally requires a minimum of three washes.

10. Refeed the cells with maintenance medium and place at 37°C for 48–72 h when the cells are ready to be harvested and assayed for CAT and luciferase activity.

[a] 2 × HBS consists of 10 g/litre Hepes (tissue–culture grade), 16 g/litre NaCl. Adjust pH accurately to 7.1 (+/−0.05) with sodium hydroxide and filter sterilize.

The following protocol describes transfection of COS-1 cells by electroporation. This technique may be applied to many different types of cells but, as already described, the precise conditions used must be determined empirically (13).

Protocol 2. Electroporation of COS-1 cells

1. Harvest cells at 70% confluence by scraping and resuspend in phosphate-buffered saline (PBS) at 8×10^6 cells/ml.

2. Pipette 0.8 ml of the cell suspension containing 6.4×10^6 cells into a Biorad Gene Pulsar Cuvette. Add 20 µl of a 1 mg/ml solution of DNA (20 µg) and place on ice for 5 min.

3. Electroporate at 450 V and 250 µF using a Biorad Gene Pulsar apparatus.

4. Place the cuvette on ice for 5 min.

5. Plate the cells from the cuvette in 4×6 cm dishes and leave at 37°C for 48 to 60 h until ready for harvesting and assaying.

Protocol 3. Harvesting of cells and analysis of luciferase activity and CAT activity

1. Wash the transfected cells twice with 4 ml of PBS at room temperature, carefully removing all the buffer from the final wash by tilting the plate.

2. Pipette 100 µl of lysis buffer[a] on to the cell monolayer of a 6 cm plate and leave the plate horizontal for 2 min until the cells have lysed and only intact nuclei are visible under the microscope.

3. Transfer the lysate to a pre-cooled microfuge tube with a Gilson pipette and store on ice. Spin for 1 min in a microcentrifuge to pellet cell debris, and transfer the supernatant to a clean tube. Store on ice.

4. To measure the activity of the internal control, add 20 µl of extract to 350 µl luciferase reaction buffer[b] in a polystyrene luminometer cuvette.

5. Load samples into the LKB 1251 luminometer and inject 33 µl of 3 mM luciferin using an LKB 1291 dispenser. Control samples from mock-transfected cells give peak activities of 0.6 to 0.8 units. Values for trans-fected NIH-3T3 cells range from 50 to 300 units.

[a] Lysis buffer: 0.65% Nonidet P-40 (NP40), 10 mM Tris–HCl (pH 8.0), 1 mM EDTA (pH 8.0), 150 mM NaCl.
[b] Luciferase reaction buffer; 25 mM glycylglycine (pH 7.8), 5 mM ATP (pH 8.0), 15 mM $MgSO_4$. This buffer is stored in 10 ml aliquots at $-20°C$.

The lysate produced by steps 1 to 3 of *Protocol 3* may be used directly to determine CAT activity. This is measured using either chromatography (5) or an assay which is non-chromatographic and directly quantitative (15).

Protocol 4. Analysis of CAT activity using a non-chromatographic assay

1. Heat 20 µl of extract to 68°C for 5 min.

2. Add 20 µl of 8 mM chloramphenicol.

3. Add 20 µl of 0.5 mM acetyl coenzyme A (acetyl CoA) containing 0.1 µCi of ^{14}C acetyl CoA (Amersham International, CFA 729, 54 mCi/mmol).

4. Add 10 µl lysis buffer and 30 µl 0.25 M Tris–HCl (pH 7.8).

5. Heat samples at 37°C for 1 h.

6. Add 100 µl of ice-cold ethyl acetate to each sample. If a large number of samples are being processed, do no more than eight samples at one time as ethyl acetate is extremely volatile. Vortex, and spin samples in a micro-centrifuge for 2 min. Chloramphenicol and its acetylated derivatives are soluble in ethyl acetate while the acetyl CoA remains in the aqueous layer.

7. Remove 80 µl of the upper organic layer and pipette directly into a scintillation vial containing 5 ml of liquid scintillant. Ensure that no material from the aqueous layer is transferred to the scintillation vial.

8. Add another 100 µl of ice-cold ethyl acetate to the aqueous layer and repeat the extraction process, this time transferring 100 µl of the organic layer to the scintillation vial.

Protocol 4. *Continued*

9. Count the samples in a scintillation counter. Mock-transfected cells give an assay background of approximately 250 c.p.m. Complete conversion gives a value of about 150000 c.p.m., but the assay is only linear up to 60000 c.p.m. Extracts which give higher values than this must be diluted in lysis buffer and re-assayed within the linear range. The data obtained from an analysis of the CAT activity in transfected cells are normalized using the results measured for the internal control (luciferase) derived from the same extracts. The final results may be presented either graphically or in a tabular form.

---

3. The use of *in vitro* translation systems and overexpression systems to analyse transcription factor function

3.1 Synthesis of proteins using coupled *in vitro* transcription and translation

A number of the functional properties of a cloned transcription factor may be analysed using *in vitro* synthesized protein. In the case of the mouse oestrogen receptor, a coupled transcription and translation system has been used to synthesize small amounts of protein for investigating DNA- and ligand-binding activity. An *in vitro* synthesis system is also particularly suitable for screening the large number of different proteins generated in the process of mutational analysis of a transcription factor. A number of plasmid vectors are available for the synthesis of RNA for the cloned factor using *in vitro* transcription. The vectors consist of:

(a) a cloning site or multiple cloning site for the insertion of the cloned cDNA

(b) promoter element(s) for RNA polymerase(s)

(c) an origin of replication and an antibiotic resistance gene for the growth and maintenance of the vector in bacterial cells

A number of bacteriophage RNA polymerases have been purified and are available for *in vitro* transcription systems; these include SP6 RNA polymerase, T3 RNA polymerase, and T7 RNA polymerase. A suitable vector, therefore, will contain one or two promoter regions for these enzymes flanking a multiple cloning site into which the coding region of the cDNA is inserted. The sequence between the start site of transcription in the promoter and the beginning of the coding region should not contain other translation initiation sites or open reading frames which may interfere with subsequent translation of the RNA. It is therefore advisable that the distance between the start site and the first codon is small, approximately 20–50 nucleotides. The efficiency of translation may also be improved by inclusion of a Kozak consensus sequence (16). A unique restriction enzyme site must be available at the 3' end of the coding sequence to linearize the vector to ensure that run-off transcripts of defined length are produced which contain the full-length coding sequence and which lack vector sequences.

The following example describes a method for producing sufficient quantities of specific RNA transcripts for the mouse oestrogen receptor to prime *in vitro* translation systems. The cDNA for the receptor is cloned into the polylinker region of the *in vitro* expression vector pSP65 (17) to generate the plasmid pMOR100 (*Figure 3*). The template DNA to be used for *in vitro* RNA synthesis is prepared by linearization of pMOR100 with the unique restriction enzyme *Hind*III. After digestion, the DNA is extracted with phenol/chloroform, recovered by ethanol precipitation and resuspended in sterile distilled water at a concentration of 1 mg/ml. An *in vitro* transcription

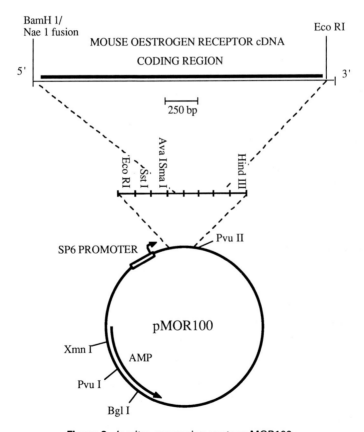

Figure 3. *In vitro* expression vector pMOR100.

reaction using SP6 polymerase and including the RNA cap structure analogue $m^7G(5')ppp(5')G$ is then used to generate capped RNA, which may be used directly in either a rabbit reticulocyte lysate or a wheat germ lysate *in vitro* translation system. Capping promotes translation *in vitro* at the level of initiation and increases the stability of the RNA transcripts. To synthesize the large amounts of RNA necessary for *in vitro* translation, a minimum of 5 μg of plasmid is required. When used in the *in vitro* transcription reaction, the yield of RNA should be between 5 and 10 μg/μg plasmid DNA template.

Protocol 5. *In vitro* synthesis of RNA

The components of the reaction are added in the order given, and the mixture is kept at room temperature to prevent the precipitation of the DNA template which may occur at 4°C in the presence of spermidine.

1. Pipette 20 μl 5 × transcription buffer[a] into a sterile microcentrifuge tube.

2. Add 10 μl 100 mM dithiothreitol (DTT).

3. Add 100 units of human placental ribonuclease inhibitor.

4. Add 20 μl 5 × CAP buffer.[b]

5. Add sufficient RNase-free distilled water so that the final reaction volume will be 100 μl.

6. Add 5 μl linearized DNA template at 1 mg/ml.

7. Add 50 units SP6 RNA polymerase.

8. Incubate at 37–40°C for 60 to 90 min.

9. Add RQ1 RNase-free DNase (Promega) to a concentration of 1 unit/μg DNA, and incubate at 37°C for 15 min to remove the DNA template.

10. Add an equal volume of a 50:50 mixture of phenol:chloroform, vortex, spin for 2 min in a microcentrifuge and remove the aqueous layer to a clean tube. Precipitate the RNA by addition of 100 μl 5 M ammonium acetate (pH 5.4) and 400 μl ethanol. Place on dry ice for 30 min and spin for 10 min in a microcentrifuge. Wash the RNA pellet with 200 μl 70% ethanol, spin for 10 min in a microcentrifuge, remove the supernatant and allow the pellet to dry. Resuspend the RNA in 20 μl RNase-free distilled water and store at −70°C. The RNA is quantitated by measurement of the A_{260}. (A 1 mg/ml solution of RNA has an A_{260} of 25.)

[a] 5 × transcription buffer (store at −20°C): 200 mM Tris–HCl (pH 8.0), 30 mM MgCl, 10 mM spermidine.
[b] 5 × CAP buffer (store at −20°C): 2.5 mM ATP, UTP, and CTP, 250 μM GTP, 2.5 mM $m^7G(5')ppp(5')G$.

Protocol 6. Synthesis of protein by *in vitro* translation using a rabbit reticulocyte lysate

1. Thaw an aliquot of lysate (200 μl) on ice and prepare a cocktail by adding:
   - 4 μl of a 1 mM amino acid mixture minus methionine
   - 2 μl of 10 mM ZnCl₂ to give a final concentration of 100 μM

   The supplement of the lysate with zinc is specific for translation of the oestrogen receptor which requires zinc for the structural organization of the DNA-binding domain.

2. Prepare 0.5 to 1.0 μg of *in vitro* synthesized mRNA in a total volume of 10 μl distilled in a microcentrifuge tube and add 20 μl of the lysate cocktail. Mix gently.

3. Transfer 9 μl of the translation mix into a clean microcentrifuge tube and add 1 μl L-[³⁵S]methionine (1000 Ci/mmol) at 15 mCi/ml to give a final [³⁵S]methionine concentration of 1.5 mCi/ml.

4. Add 1 μl of 1 mM unlabelled methionine to the remaining translation mix.

5. Incubate the labelled and unlabelled translation reactions at 30°C for 60 min.

6. Add glycerol from a 50% stock to give a final concentration of 15% and store the translations at −70°C.

Translated proteins labelled with [³⁵S]methionine are used to assess the size, purity, and yield of the protein by SDS-polyacrylamide gel electrophoresis. Unlabelled protein can be used for DNA-binding assays and, in the case of the oestrogen receptor, for hormone-binding studies. Protein can also be synthesized *in vitro* using a wheat-germ lysate, although the yield from this system is generally lower than from a rabbit reticulocyte lysate, especially for proteins of greater than 40 kDa which are translated with reduced efficiency. Although suitable for the rapid production and analysis of a large number of different samples, *in vitro* transcription and translation systems have a number of disadvantages. In particular, the quantity of protein produced is small, and a large proportion of the synthesized protein may be incorrectly folded. Estimates indicate that less than 5% of the receptor protein produced in a reticulocyte lysate is folded correctly. The protein may also lack normal post-translational modifications such as phosphorylation and glycosylation. The suitability of coupled transcription and translation systems for general analysis of transcription factor function must, therefore, be determined empirically. The methods described above have been used successfully to produce oestrogen receptor protein in sufficient quantities to study both the DNA- and ligand-binding properties of the receptor. However, similar *in vitro* expression techniques have not been successful in the analysis of the DNA-binding activity of the glucocorticoid, androgen, and progesterone receptors.

3.2 Overexpression in insect cells following infection with baculovirus vectors

An alternative source of protein is from overexpression in insect cells following infection with a recombinant baculovirus vector into which has been inserted a cDNA clone of the protein under investigation. This type of approach is used to produce large quantities of protein suitable for purification and structural analyses. Detailed methods for the maintenance of insect cells and the manipulation of baculovirus vectors have been described by a number of groups (18, 19). In studying the functional properties of a transcription factor, one major advantage of synthesizing large amounts of material is that sufficient quantities are made available to add to *in vitro* transcription reactions allowing a direct analysis of transcriptional activation (see Chapter 3). The large amounts of protein produced may also be used for biochemical studies such as phosphorylation and for structural studies using biophysical techniques. Very high levels of expression, however, may also result in the protein becoming insoluble and, therefore, unsuitable for further analysis.

The mouse oestrogen receptor has been expressed in *Spodoptera frugiperda* cells (Sf9) by use of a baculovirus vector system (20). Mutant forms of the receptors have also been introduced into the baculovirus overexpression system by the vectors and methods described by Kitts *et al.* (21). The receptor produced has been used in DNA- and ligand-binding studies and has also been used to supplement Hela cell extracts in an *in vitro* transcription assay. The overexpressed protein appears to be correctly folded as judged by ligand- and DNA-binding properties. However, although the addition of the receptor protein to an *in vitro* transcription assay results in an increase in the basal level of transcription (22), hormone-dependent transactivation has still to be demonstrated. This may reflect differences in the modification of the receptor expressed in insect cells compared with the protein synthesized in a mammalian cell system.

3.3 Overexpression in mammalian cells

Transcription factors may be overexpressed in mammalian cells by using the COS-1 cell system (23), although the levels of protein produced may be significantly lower than with the baculovirus expression system. Recombinant vectors containing a SV40 early promoter and origin of replication are amplified in COS-1 cells, allowing high levels of expression of mRNA and protein. This method is simple to perform and is suitable for the analysis of a large number of mutant forms of a protein; however, it is not suitable for sustained high levels of production of the protein. If this is required, or if it is necessary to express the cloned factor in specialized cell types, then a different vector system must be used. Other eucaryotic viral expression systems to consider are those based on vaccinia, bovine papillomavirus, or retroviruses (3). The expression of the mouse oestrogen receptor following transient transfection by electroporation into COS-1 cells has been described in the first part of this chapter in *Protocol 2*. Transient transfection results in up to 40% of cells expressing protein, as shown by immunoflourescence with specific antisera. A method for the preparation of a whole-cell extract containing undegraded protein is described below in *Protocol 7*, in which cells are lysed in a buffer containing a cocktail of protease inhibitors. This method is also suitable for the preparation of extracts from Sf9 cells infected with a recombinant baculovirus vector.

Protocol 7. Preparation of a whole-cell extract

1. Remove the medium from the cell monolayer and wash the cells three times with ice-cold PBS. Scrape the cells from the dish and transfer the cell suspension in PBS into a sterile plastic universal. Recover the cells by centrifugation at $2000 \times g$ for 5 min. Cell pellets may be stored at $-70\,°C$. Infected Sf9 cells are loosely attached to culture flasks; therefore, the cells are harvested by shaking the flask and the cells are then recovered from the medium by centrifugation at $2000 \times g$ for 10 min. Wash the cell pellet three times with ice-cold PBS. The pellet may be stored at $-70\,°C$ until required.
2. Resuspend the cell pellet in 10 volumes of $1 \times$ high salt buffer.[a]
3. Pass the lysed cell suspension through a 25-gauge needle five times.
4. Centrifuge at $50000 \times g$ for 15 min to pellet insoluble debris.
5. Store the supernatant in aliquots at $-70\,°C$.

[a] $1 \times$ high salt buffer: 0.4 M KCl, 20 mM Hepes (pH 7.4), 1 mM DTT, 20% glycerol, 0.5 mg/ml bacitracin, 40 μg/ml phenylmethylsulphonyl fluoride (PMSF), 5 μg/ml pepstatin, 5 mg/ml leupeptin.

3.4 Comparison of expression systems

In summary, no single method of expression is ideally suitable for the detailed analysis of a cloned factor. *In vitro* transcription and translation systems allow the rapid analysis of either a large number of proteins or an extensive series of mutant forms of a single protein. *In vitro* synthesized protein, however, is produced in small amounts and may lack normal processing and modification. Expression in bacteria may be used to synthesize large quantities of protein which may then be analysed in functional assays or used for the determination of protein structure. The protein produced, however, may lack authentic processing and other post-translational modifications such as glycosylation and phosphorylation. This problem may be overcome by expression in eucaryotic cells. The baculovirus overexpression system allows the production of large amounts of material, but in view of the difficulty of isolating recombinant virus, it is not suitable for the analysis of a large number of different proteins. Finally, expression in mammalian cells by use of vectors based on viral systems is most likely to produce correctly processed or modified protein, although high levels of expression may be restricted to specialized cell types or require the isolation of stable cell lines. A detailed analysis of a cloned factor will, therefore, require that a number of different expression systems are used.

4. Analysis of functional domains by deletion mutagenesis

4.1 Introduction

The organization of functional domains within a transcription factor may be determined by deletion mutagenesis. This involves either progressive deletion of sequences from the N- or C-terminus of the protein, or the construction of a series of internal deletions in which small regions of the sequence are removed. Deletion analysis determines the positions of the boundaries of individual domains, which then allows the analysis of a particular functional property to be investigated, either in isolation or as part of a chimeric protein. This section will describe the construction of a series of deletion mutants of the cloned transcription factor, which will be followed by a method using a gel retardation assay to analyse the DNA-binding activity of the protein. We will then discuss the mapping of the boundaries of functional domains for both DNA binding and dimerization by a combination of deletion mutagenesis and the gel retardation assay to investigate the formation of heterodimeric forms of the protein. The final part of this section will describe the use of deletion mutants in the analysis of other functional properties, including ligand binding and transcriptional activation.

4.2 Construction of mutants by deletion mutagenesis

A range of methods have been devised for the generation of a series of deletion mutants of a protein from a cloned cDNA. These include:

(a) The use of specific exonucleases to remove sequences progressively from the N- and C-terminal regions, generating at random a range of different sizes of clones.

(b) Specific deletion using internal restriction enzyme sites.

(c) The formation of deletions by the polymerase chain reaction (PCR).

In all cases, the precise position of the deletion made must be verified by sequence analysis. Deletion of N-terminal sequences results in removal of the initiator codon which must subsequently be replaced, while C-terminal deletions require the insertion of a termination codon.

4.2.1 Deletion mutants of the mouse oestrogen receptor

A series of N- and C-terminal deletion mutants of the mouse oestrogen receptor (*Figure 4*) have been constructed using both internal restriction enzymes in the cDNA clone and progressive deletion with Bal-31 nuclease. The resulting cDNA clones were inserted into the polylinker region of the *in vitro* expression vector pSP64 to allow synthesis of mutant forms of the receptor protein by coupled *in vitro* transcription and translation (*Protocols 5 and 6*). The sequence surrounding the initiator codon of the mouse oestrogen receptor is similar to the Kozak consensus sequence for efficient translational initiation. All N-terminal deletion mutants were, therefore, constructed to include the first three amino acids of the normal receptor protein to ensure correct and efficient initiation of translation.

4.2.2 Internal deletions

Internal deletion analysis may be used systematically to determine the positions of functional domains. This approach has been used in the mapping of regions in the human oestrogen receptor (25). An example of the use of internal deletion analysis is the identification of sequences required for nuclear localization. A number of transcription factors contain a short sequence of mainly basic amino acids which form a nuclear targeting

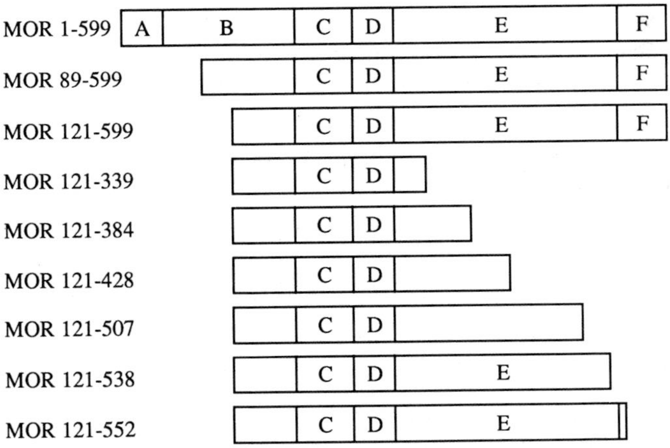

Figure 4. Deletion mutants of the mouse oestrogen receptor. The numbers refer to the amino acid sequence of the receptor protein. The regions labelled A–F are derived from sequence comparisons between the oestrogen and glucocorticoid receptors (24). Region C is the DNA-binding domain and region E is the ligand-binding domain.

sequence that functions as a nuclear localization signal (26). Internal deletion analysis has been used to identify this type of functional element in many transcription factor proteins. It has been demonstrated that deletion of four amino acids between residues 638 and 642 in the progesterone receptor generates a protein which is predominantly cytoplasmic in its intracellular distribution. These forms of the receptor have been used to investigate nucleocytoplasmic shuttling (27) and dimerization (28).

4.3 Analysis of DNA-binding activity by a gel retardation assay

Methods have been described in the previous section for the preparation of whole-cell extracts. The DNA-binding activity in extracts is determined using a gel retardation assay in which a ^{32}P-labelled oligonucleotide containing a response element or binding site for the transcription factor is first incubated with the whole-cell extract. The protein–DNA complexes which form are separated from the unbound or free DNA on vertical acrylamide gels and visualized by autoradiography. The single-stranded oligonucleotides forming the oestrogen receptor binding site have the sequence 5'-CTAGAAAGTCA-GGTCACAGTGACCTGATCAAT-3' and 5'-CTAGATTGATCAGGTC-ACTGTGACCTGACTTT-3'. These are annealed and labelled with [^{32}P]dCTP, using the Klenow fragment of DNA polymerase to form a double-stranded probe containing a consensus oestrogen receptor binding site.

Protocol 8. Preparation of ^{32}P-labelled oligonucleotide for DNA-binding assays

1. Anneal the DNA to form the probe by mixing in a microcentrifuge tube 0.5 μg of each oligonucleotide in a total volume of 100 μl of 0.1 M NaCl, 10 mM Tris–HCl (pH 8.0), and 1 mM EDTA.

2. Heat the sample to 85°C for 3 min in a small water bath, switch off the bath and allow the sample to cool for 30 min to below 45°C.

3. Add 200 μl of ethanol, vortex, and place the sample on dry ice for 30 min to precipitate the DNA.

4. Spin in a microcentrifuge for 15 min, carefully remove the supernatant and wash the pellet with cold 70% ethanol. Spin for 15 min, remove the supernatant and allow the pellet to dry.

5. Dissolve the annealed oligonucleotides in 10 μl distilled water to give a concentration of 100 ng/μl.

6. Label the DNA using the following reaction:
   - 200 ng annealed DNA   2 μl
   - 10 × repair buffer (0.5 M Tris–HCl (pH 7.6), 0.1 M MgCl$_2$)   2 μl

- distilled water   10 μl
- [^{32}P]dCTP in aqueous solution at 3000 Ci/mmol and 10 mCi/ml
  (Amersham)   5 μl
- Klenow DNA polymerase   1 μl

Incubate at 37°C for 30 min.

7. Add an equal volume of 50:50 mixture of phenol/chloroform, vortex, spin
   for 2 min in a microcentrifuge and remove the upper aqueous layer to a
   clean tube. Repeat the phenol/chloroform extraction and remove the
   aqueous layer to a clean tube. Add 20 μl 5 M ammonium acetate and 80 μl
   ethanol. Place on dry ice for 30 min to precipitate and recover the DNA by
   centrifugation in a microcentrifuge for 15 min.

8. Resuspend the pellet in 20 μl of distilled water. Add 20 μl 5 M ammonium
   acetate and 80 μl ethanol. Place on dry ice for 30 min to precipitate and
   recover the DNA by centrifugation as before.

9. Wash the labelled DNA probe with 70% ethanol and centrifuge for
   15 min. Carefully remove the supernatant and leave the pellet to dry.
   Resuspend the DNA in 20 μl of distilled water to give a final concentration
   of 10 ng/μl.

Protocol 9. Analysis of DNA-binding activity by a gel retardation assay

1. Pipette 10 μl of 2 × binding buffer[a] into a microcentrifuge tube.

2. Add 1 μl of 1 mg/ml poly(dI-dC)·(dI-dC) and 1 μl of 10% bovine serum
   albumin (BSA).

3. Add lysate or whole-cell extract. The amount of protein added is determined empirically; in the example of the oestrogen receptor, suitable
   values are 2 μl of *in vitro* translated receptor, 2 μl of COS-1 cell extract, or
   the equivalent of 0.1 μl of extract from recombinant baculovirus-infected
   insect cells. To measure non-specific binding, prepare a parallel set of
   reactions using equivalent amounts of either control lysate or whole-cell
   extract lacking receptor.

4. Make the total volume up to 19 μl with distilled water and incubate the
   samples at room temperature for 15 min.

5. Add 1 μl of ^{32}P-labelled oligonucleotide at 1 ng/μl to each sample and
   incubate for 30 min at room temperature and then for 30 min at 4°C.

6. Prepare a 6% polyacrylamide gel (30% acrylamide, 0.8% bisacrylamide)
   in 0.5 × TBE buffer.[b]

7. Before running the gel, rinse the slots with 0.5 × TBE buffer to remove
   unpolymerized acrylamide. Pre-run the gel in 0.5 × TBE at 100 V for
   30 min.

Protocol 9. *Continued*

8. Load the samples directly on to the gel and run at 250 V. Load marker dye in the outside track(s) to monitor the electrophoresis.

9. Fix the gel in 10% acetic acid, 30% methanol for 15 min, dry and auto-radiograph.

ᵃ 2 × binding buffer: 40 mM Hepes (pH 7.4), 100 mM KCl, 2 mM 2-mercaptoethanol, 20% glycerol.
ᵇ 1 × TBE 0.09 M Tris–borate (pH 8.3), 2 mM EDTA.

4.3.1 Application of the gel retardation assay to transcription factor analysis

The gel retardation assay may be applied to determine the affinity of the protein–DNA interaction by using increasing concentrations of input DNA probe and measuring the amount of the protein–DNA complex formed with an Ambis β-scanner. To compare directly either different samples or different reaction conditions, it is important that the DNA probe is in excess. Specific binding of the protein to DNA is confirmed by the inclusion of specific antisera. This may either generate a decrease in mobility (or 'supershift') of the protein–DNA complex in the gel as a result of the binding of the antibody or may result in the loss of the complex due to interference with DNA binding (see Chapter 1). In the analysis of the functional properties of the oestrogen receptor, the gel retardation assay may be used to observe the effect of the addition of ligand on DNA binding. Hormone agonists, e.g. oestradiol, or antagonists, e.g. tamoxifen, may be added to the binding reaction either before or after the addition of the DNA probe. *Figure 5* is an autoradiograph of a gel retardation assay of a series of deletion mutants of the mouse oestrogen receptor. The positions of the free or unbound DNA probe and the retarded protein–DNA complexes are shown.

This analysis indicates that the C-terminal boundary of residues required for high-affinity DNA binding lies between amino acids 507 and 538. This region, however, is not required for specific recognition of DNA by the receptor. The gel retardation assay, together with comparison of the sequence of the mouse oestrogen receptor (29) with that of the human oestrogen receptor and other members of the steroid hormone receptor superfamily, demonstrates that the residues involved in specific interaction with DNA occur within a 66 amino acid domain. This corresponds to amino acids 188 to 252 in the mouse oestrogen receptor. Biophysical studies (30) have demonstrated that the DNA-binding domain of the human oestrogen receptor is contained within a small region between amino acids 180 and 262. The C-terminal boundary for high-affinity DNA binding mapped between residues 507 and 538 may, therefore, indicate the position of residues required for protein–protein interactions between receptors which are required for maintaining the stability of the protein–DNA complex.

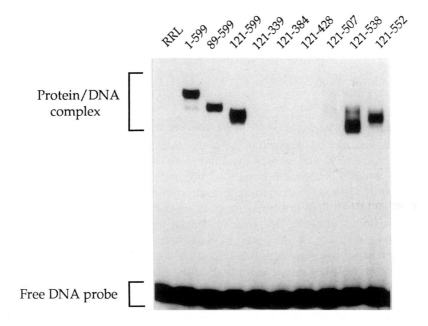

Figure 5. Gel retardation assay of oestrogen receptor deletion mutants. The numbers refer to the deleted receptor proteins shown in *Figure 4*.

4.4 The analysis of transcription factor dimerization

4.4.1 Introduction

Interaction between transcription factors provide a mechanism for generating the large range of control systems required for different biological processes. The ability of a particular factor to bind to a response element may depend not only on the sequence of the DNA but also on protein–protein interactions, which include the formation of homodimers or heterodimers with other related proteins (31). Transcription factors which bind to DNA in the form of a dimer include those where the dimerization interface has been shown to be mediated either by a heptad repeat of leucine residues forming a coiled-coil structure of parallel α-helices termed a leucine zipper (32) or those which interact through a helix–loop–helix motif (33). The AP1 transcription factor complex is formed by a heterodimeric association of two different proteins, Fos and Jun (34). In both Fos and Jun the leucine zippers are adjacent to a region of basic amino acids which directly interact with DNA. These proteins, which are members of a large family of closely related transcription factors and other proteins with related structure, are termed basic zipper (bZIP) proteins. In contrast to the bZIP proteins, the amino acids identified as being essential for high-affinity DNA binding in the C-terminal region of the mouse

163

oestrogen receptor are approximately 250 amino acids away from the region involved in specific contacts with DNA. This implies, therefore, that the steroid receptors may represent a class of transcription factor with a novel dimerization activity and that residues in the hormone-binding domain may be involved directly in protein–protein contacts which form a major part of the dimer interface.

4.4.2 Analysis of dimerization by the gel retardation assay

Dimerization may be analysed in the presence of DNA by the gel retardation assay (*Protocol 9*) and is measured indirectly by the formation of hetero-dimeric complexes between proteins of different size. Co-translation of a full-length protein and a deletion mutant results in the formation of a heterodimer which migrates to an intermediate position in the gel relative to the two different homodimers of the protein. This type of analysis is illustrated in *Figure 6* by an assay in which the full-length oestrogen receptor MOR 1-599 is co-translated with an N-terminal deletion mutant MOR 121-599, both of which retain residues required for high-affinity DNA binding. Although this assay is measuring a dimerization function, detection of the protein–DNA complexes depends on the ability of the proteins to bind to DNA. A method for analysing dimerization directly in the absence of DNA by immunoprecipitation is provided in *Protocol 11*.

4.5 Analysis of the boundaries of functional domains for ligand binding and transcriptional activation

The series of deletion mutants described above may also be used to determine the boundaries of functional domains required for transcriptional activation and, in the case of the oestrogen receptor, hormone binding. The extent of the hormone-binding domain is determined by a ligand-binding assay using *in vitro* translated receptors. The ligand-binding assay can also be used to determine the relative amounts of receptor protein in a whole-cell extract (*Protocol 8*) or to determine the dissociation constant (K_d) of the receptor for ligand. To determine a K_d for ligand binding, a range of concentrations of input steroid is used and the K_d for the oestrogen receptor has been determined to be 0.1 nM. The C-terminal boundary for ligand binding lies between amino acids 507 and 538 (35).

Protocol 10. Ligand-binding assay

1. Prepare four samples, each containing 5 μl of *in vitro* translated receptor lysate in a total volume of 50 μl of 10 mM Tris–HCl (pH 7.4), 1 mM EDTA, and 1 mg/ml BSA.

2. Add 16α-[^{125}I]iodoestradiol to a concentration of 10 nM.

3. To determine non-specific binding, add the competitor diethylstilbestrol to a final concentration of 1 mM to two of the samples.

4. Incubate at 4°C overnight.

5. Prepare a DCC suspension as follows:

- dextran T70   0.1 g
- Norit A charcoal (Sigma)   1.0 g
- 1 M Tris–HCl (pH 7.4)   4.0 ml
- 0.5 M EDTA   0.8 ml

Make up to 400 ml with distilled water and store at 4°C.

6. Add 50 µl of DCC suspension to each sample and incubate at 4°C for 5 min.

7. Spin for 5 min in a microcentrifuge, remove 90 µl of the supernatant and count the bound fraction in a γ-counter.

8. Calculate the unbound or free ligand by resuspending the pellet in 10 µl of distilled water and counting.

The boundaries of transactivation domains are determined by transfer of the deleted cDNA clones into a eucaryotic expression vector and the introduction of the mutant proteins into mammalian cells using a co-transfection assay as has been described in Section 2.4. Transfer of the series of deletion mutants of the mouse oestrogen receptor into pJ3Ω and co-transfection assays (*Protocols 2* and *4*) have been used to demonstrate that the C-terminal boundary of a hormone-inducible transactivation domain in the mouse oestrogen receptor occurs between amino acids 538 and 552 (36).

4.6 Summary

As has been demonstrated by the above examples, deletion analysis of a cDNA clone allows the determination of a functional boundary within a protein. A series of deletion mutants may also be used to analyse a number of different functional properties. When the boundaries of functional domains have been defined, a more detailed analysis of individual domains may be carried out using chimeric proteins.

5. Mapping of functional domains with chimeric proteins

5.1 Introduction

Chimeric proteins are formed by the joining together of separate functional domains from different proteins. In practice, this may be accomplished by the ligation of cDNA clones, or parts of cDNA clones, producing recombinant

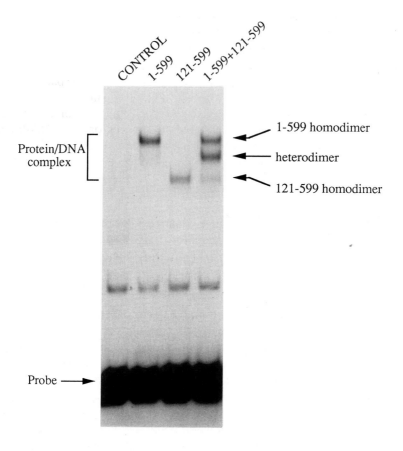

Figure 6. Analysis of protein dimerization by the gel retardation assay.

molecules which on transcription and translation form fusion proteins from a single open reading frame. The ability to form chimeric proteins by the swapping of functional domains between transcription factors was originally determined by experiments in which the DNA-binding domain of the bacterial repressor LexA was fused to the activation domain of GAL4 to generate a transactivator which operated through a LexA-binding site (37). Chimeric proteins have been mainly used to identify and characterize transcriptional activation domains in transcription factors. This involves the fusion of a putative transactivation region from one protein to a DNA-binding domain from a different protein for which the DNA response element has been well characterized. From this type of experiment, it has been concluded that transcription factors are modular in organization and this has been confirmed by studies using a large number of transcription factors, including the oestrogen and glucocorticoid receptors. Domains may be interchanged between the receptors to switch DNA binding, ligand binding, and transcriptional activation functions. The relative positions of these domains within the chimeric proteins may be varied, implying that each domain acts as a separate module and represents an independent functional unit. Domain swap experiments have also been used to analyse the dimerization function of transcription factors. The interchange of leucine zipper sequences or the DNA-binding domains of the bZIP proteins Fos and Jun has demonstrated the relative importance of heterodimerization and homodimerization in the formation of a stable dimer interface (38).

5.2 Identification of a transactivation domain with a chimeric protein

The following example describes the identification of a transactivation domain by construction of a chimeric receptor consisting of the DNA-binding domain of the yeast protein GAL4 linked to the hormone-binding domain of the mouse glucocorticoid receptor. The eucaryotic expression plasmid pGAL4-GR has been described by Danielian *et al.* (37). This consists of a 975-nucleotide *Eco*R1 to *Xba*1 fragment containing the C-terminal coding region of the mouse glucocorticoid receptor from amino acids 506 to 783, with 144 nucleotides of 3′ untranslated sequence linked to the 147 amino acid DNA-binding domain of GAL4 in the vector pSG424 (40). To construct a fusion protein, the sequence of pSG424 was modified using an oligonucleotide cassette (see Section 6.3) so that the *Eco*R1 site in the polylinker was removed and replaced with another *Eco*R1 site in the same reading frame as the site in the receptor. A diagram of the vector pGAL4-GR and the sequence of the olignucleotides used is shown in *Figure 7*.

Transactivation by the chimeric receptor is determined using the co-transfection assay as described in *Protocols 1, 3,* and *4*. The expression vector pGAL4-GR is introduced into NIH-3T3 cells by calcium phosphate co-precipitation, together with the reporter plasmid pG5E1BCAT (41) which consists of five GAL4DNA-binding sites linked to the E1b promoter and CAT gene. The transcriptional activity is determined by measurement of CAT activity as has been described in *Protocol 4*. The combined approaches of chimeric proteins and deletion analysis, as described in the previous section, allow the extent of a functional transactivation domain to be determined. Point mutagenesis may subsequently be used to determine residues directly involved in transcriptional activation.

In summary, deletion mutagenesis may be used to determine the boundaries of functional domains. The analysis of domains in chimeric proteins allows the confirmation of these boundaries and demonstrates the extent to which individual domains can function as separate units.

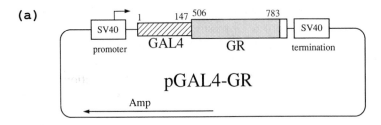

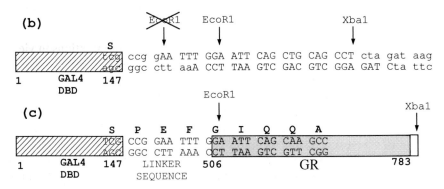

Figure 7. (a) The chimeric protein expression vector pGAL4-GR. (b) Modified sequence of the GAL4 DNA-binding domain expression vector pSG424, in which the polylinker region between the *Eco*R1 site and *Xba*1 site has been replaced with an oligonucleotide cassette. The original *Eco*R1 site has been destroyed and another site introduced in frame with the *Eco*R1 site in the mouse glucocorticoid receptor at amino acid 506. The sequence of the vector pSG424 is shown by lower-case letters and the oligonucleotide cassette by upper-case letters. (c) Sequence of the junction of the fusion protein GAL4-GR.

6. Analysis of functional domains by point mutagenesis

6.1 Introduction

The identification of the boundaries of a domain within a transcription factor allows more specific approaches, such as point mutagenesis, to be used to identify individual amino acid residues required for functional activity. The decision as to which changes to introduce, however, may be unclear, particularly when the functional domain is large. A number of approaches should be considered to assist in the selection of individual residues to mutate.

(a) If the functional activity has been identified by deletion mutagenesis, the important residues may occur close to the N- or C-terminal boundaries.

(b) A comparison of the primary amino acid sequence with the sequence of other highly related proteins may reveal conserved or homologous regions. Conservation of sequence may indicate the position of essential residues required for functional activity.

(c) A comparison should also be made of the primary amino acid sequence with that of unrelated proteins known to encode a similar function.

(d) Conserved motifs may be identified which indicate the positions of potential sites for modification of the protein by processes such as phosphorylation.

(e) Analysis of the character of the individual amino acids, i.e. whether hydrophobic, uncharged, or charged, can reveal specific residues to be selected as targets for mutagenesis and, therefore, determine the type of mutation to introduce.

It is important to consider the effect of the introduction of mutations in view of the fact that alteration of a particular amino acid may result in a major structural change in the protein and, therefore, destroy functional activity. If the overall structure of the protein is perturbed, then the individual residue mutated may itself not be directly involved in transcription factor function. The choice of an amino acid such as alanine, which is neutral in character, may avoid this problem to some extent. One major advantage of the nuclear hormone receptors is that a number of different functional properties are localized in the hormone-binding domain and, therefore, the ability to alter one function without affecting the others implies that the overall structure of the protein is unaffected. The DNA-binding activity of a number of transcription factors has been investigated using point mutagenesis. The amino acids required for specific interaction with the DNA-binding site have been determined for the glucocorticoid and oestrogen receptors and these results have been subsequently confirmed by structural analysis. Point mutagenesis of individual residues and the interchange of these residues between the two proteins has identified three amino acids at the base of the first zinc finger in the DNA-binding domain and at the start of the interfinger region. These three amino acids confer the ability of the proteins to discriminate between a glucocorticoid response element or an oestrogen response element (42).

6.1.1 Identification of individual residues involved in dimerization

The approaches listed above are illustrated by using as an example the identification of amino acids implicated in the formation of a major part of the dimerization interface of the mouse oestrogen receptor. As has been shown by deletion analysis, sequences near the C-terminus of the receptor protein are required for high-affinity DNA binding in the gel retardation assay, with the C-terminal boundary of this activity being between amino acids 507 and 538 (see Section 4.3.1). A sequence comparison of this region of the receptor with other highly related proteins of the steroid and nuclear hormone receptor superfamily indicates a conserved region (*Figure 8*). When a consensus

```
                          511           518           525           532
mOR   505-535  H R R L A Q L L L I L S H I R H M S N K G M E H L Y N M K C K
hER   501-531  Q Q R L A Q L L L I L S H I R H M S N K G M E H L Y S M K C K
hAR   852-882  S R R F Y Q L T K L L D S V Q P I A R E L H Q F T F D L L I K
hPR   867-897  S Q R F Y Q L T K L L D N L H D L V K Q L H L Y C L N T F I Q
hGR   712-745  W Q R F Y Q L T K L L D S M H E V V E N L L S Y C F Q T F L D
hRARα 341-371  P H M F P K M L M K I T D L R S I S A K G A E R V I T L L M E
hRXRα 412-438  P G R F A K L L L R L P A L R S I G L K C L E H L F F F K L I
hTRα  359-389  T H F W P K L L M K V T D L R M I G A C H A S R F L H M K V E
hVDR  347-377  H L L Y A K M I Q K L A D L R S L N E E H S K Q Y R C L S F Q

                           L
                 R F X Q L - X L L X X L R
                           T
```

Figure 8. Sequence comparison of the region of the mouse oestrogen receptor implicated in high-affinity DNA binding and dimerization. The sequences are described in reference 43 with the exception of hRXRα (44). The shaded columns indicate conserved hydrophobic amino acids, and the numbers at the top of the columns correspond to amino acids in the mouse oestrogen receptor. The sequence shown in bold is a consensus sequence derived from the steroid hormone receptors.

sequence derived from this region is compared with other proteins, a conserved heptad repeat of hydrophobic residues is revealed, with homology to the coiled coil structure of the muscle proteins myosin and tropomyosin and with the leucine zipper sequence which has been implicated in dimerization of a number of other transcription factors. This sequence analysis implies that important amino acids to target for mutation in the mouse oestrogen receptor are those of hydrophobic character forming the key residues of the heptad repeat.

6.2 Introduction of specific point mutations by oligonucleotide-directed mutagenesis

Mutagenesis of a cloned cDNA sequence may be carried out using a number of different techniques. These are described in detail in the *Practical Approach* series volume titled *Directed Mutagenesis* (45). The three main approaches are:

(a) site-directed mutagenesis
(b) cassette mutagenesis
(c) recombination and mutagenesis of DNA sequences by PCR

Specific amino acid mutations are introduced by alteration of the DNA sequence of the cloned transcription factor by site-directed mutagenesis. This results in the change of the coding sequence and, therefore, allows synthesis of a mutated form of the protein following expression either *in vitro* or *in vivo*. The first two methods listed above have been used to introduce changes into the cDNA sequence of the mouse oestrogen receptor. Site-directed mutagenesis has been carried out using the Amersham oligonucleotide-directed *in vitro* mutagenesis system, which is based on the method described by Eckstein (46). An oligonucleotide containing the mutated sequence is annealed to a single-stranded DNA template and extended by Klenow DNA polymerase in the presence of T4 DNA ligase. During this synthesis, a thionucleotide is incorporated into the mutant strand of the double-stranded DNA. This prevents complete cleavage of the DNA by certain restriction enzymes and, therefore, during digestion of the DNA, breaks are only generated in the non-mutant strand. The nicked double-stranded DNA is then treated with exonuclease III to digest away all or part of the non-mutant strand. The mutant strand is then used as a template to produce double-stranded DNA containing the required mutation and the change confirmed by sequence analysis. The oligonucleotide-directed *in vitro* mutagenesis system has been used to mutate the amino acids in the mouse oestrogen receptor identified as key residues of the heptad repeat (leucine-511, isoleucine-518, glycine-525, and methionine-532). The following section describes the design of oligonucleotides and the approach taken to mutate the leucine at position 511 to arginine (L511-R). Detailed protocols for the individual reactions involved are available from Amersham International.

6.2.1 Site-directed mutagenesis of the mouse oestrogen receptor

A 790 nucleotide fragment from pMOR100 was isolated using the unique *Xba*I site at position 1202 and the unique *Sst*I site at position 1992 in the receptor cDNA and it was cloned into the polylinker of the phage M13mp19. Single-stranded DNA was then prepared and used as a template for oligonucleotide-directed mutagenesis. The restriction enzymes were chosen because they allow transfer of a small fragment of 522 nucleotides between pMOR100 and the M13 vector, using the *Sst*I site and a unique *Bgl*II site at position 1470. The size of the fragment should be as small as possible because the sequence of the entire region transferred must be verified following the mutagenesis reactions. The sequence of the oligonucleotide selected for mutagenesis is shown in *Figure 9* and is designed to be complementary in sequence to the single-stranded DNA isolated from the M13mp19 vector. The nucleotide generating the specific mutation are limited to the centre of the oligonucleotide sequence to ensure efficient annealing of the primer to the template DNA.

Following mutagenesis, the sequence of the *Bgl*II to *Sst*I fragment is analysed to confirm that the required mutation has been introduced and that the flanking sequences are unaltered. The mutated fragment is then transferred back into pMOR100 to form the mutated plasmid pMOR-L511-R. Coupled *in vitro* transcription and translation as described in *Protocols 5* and *6* may then be used to synthesize a receptor protein containing the required point mutation.

The above example demonstrates that for this type of method to be used a number of conditions must be satisfied. These include the requirement for suitable enzyme sites to be available flanking the region to be mutated. The

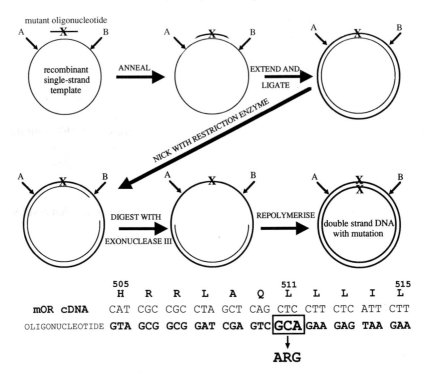

		505 H	R	R	L	A	Q	511 L	L	L	I	515 L
mOR cDNA		CAT	CGC	CGC	CTA	GCT	CAG	CTC	CTT	CTC	ATT	CTT
OLIGONUCLEOTIDE		GTA	GCG	GCG	GAT	CGA	GTC	GCA	GAA	GAG	TAA	GAA

ARG

Figure 9. Oligonucleotide-directed site mutagenesis and the sequence of the oligo-nucleotide used to change the leucine at position 511 in the mouse oestrogen receptor to arginine. A and B refer to the unique enzyme sites used for subcloning into the M13 vector. X indicates the mutation in the oligonucleotide sequence.

sites must also be compatible with sites in the multiple cloning site of a vector designed for the isolation of single-stranded DNA. Finally, the sites must be relatively close together, since the fragment transferred during mutagenesis must be sequenced to ensure that only the required mutation has been introduced.

6.3 Introduction of specific point mutations by use of oligonucleotide cassettes

A simple and easy procedure for the introduction of a large number of different mutations into a particular region of a protein is by the use of oligonucleotide cassettes. For this approach to be used, however, restriction enzyme sites must be available flanking the region to be modified between which the synthesized oligonucleotides can be inserted. If suitable restriction sites are not available, it may be possible to introduce mutations using site-directed mutagenesis to change the nucleotide sequence and generate unique sites without altering the coding potential of the sequence. The sequence identified as being a possible region involved in receptor dimerization has a unique *Cel*II site at position 1716 at the 5′ end. No suitable restriction site, however, is available 3′ of the sequence. Site-directed mutagenesis, as described above, has been used to modify the DNA sequence of the cDNA clone to create a unique restriction site without altering the coding potential of the sequence. This technique has been used to generate a unique *Kpn*1 site in the mouse oestrogen receptor at position 1801 at the C-terminal end of the conserved region. The oligonucleotide chosen to construct this mutation contains two changes from the normal receptor cDNA sequence. The sequence of the oligonucleotide and the amino acid sequence of this region of the mouse oestrogen receptor are shown in *Figure 10*. The introduction of a unique *Kpn*I site into pMOR100 generates a plasmid pMORK containing an alternative cDNA sequence, which on expression *in vitro* or *in vivo* results in the synthesis of a normal or wild-type receptor protein.

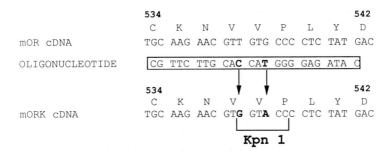

Figure 10. Generation of a *Kpn*1 site using site-directed mutagenesis without altering the coding sequence of the cDNA clone. The top line shows the sequence of the mOR cDNA between amino acids 534 and 542. The sequence of the oligonucleotide used is boxed and the modified sequence of mORK shown in the bottom line.

Specific mutations may now be introduced into this vector by the synthesis of complementary pairs of oligonucleotides with sequences compatible with a *Cel*II site at the 5′ end and a *Kpn*I site at the 3′ end. This procedure has been used to introduce single point mutations, multiple mutations, and insertion and deletion mutations between amino acids 511 and 537 in the hormone-binding domain of the mouse oestrogen receptor.

6.4 The effect of point mutations on DNA binding and ligand binding

The mutated proteins constructed using the methods described in the previous sections have been analysed for their ability to bind ligand (*Protocol 10*) and to bind to DNA with high affinity (*Protocol 9*). The effect of the single point mutations L511-R, I518-R, G525-R, and M532-R on DNA-binding

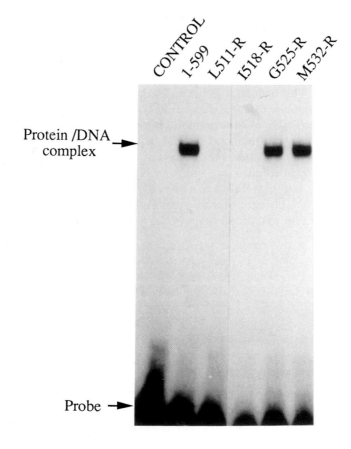

Figure 11. Gel retardation assay using *in vitro* translated oestrogen receptor proteins containing single point mutations.

activity of the mouse oestrogen receptor are shown in *Figure 11*. This demonstrates that the mutations L511-R and I518-R disrupt high-affinity DNA binding, while the mutations G525-R and M532-R have no effect.

The mutated proteins have also been analysed for their ability to bind oestradiol in the ligand-binding assay. L511-R and M532-R have $K_d = 1.1$ and $K_d = 0.56$, respectively, I518-R and G525-R, however, show no detectable binding of oestradiol. These data demonstrate that residues required for ligand binding overlap in part with those required for high-affinity DNA binding.

In the application of point mutagenesis in the analysis of transcription factor activity, the observation that mutant proteins retain one functional property while another is destroyed, as in the case of L511-R, suggests that the overall structure of this mutant protein is unaltered and that the particular amino acid changed is directly involved in high-affinity DNA binding.

6.5 Direct analysis of dimerization of transcription factors

A method for analysing dimerization using a combination of deletion muta-genesis and the gel retardation assay (*Protocol 9*) has been described in Section 4.4.2. The detection of dimeric forms of the protein in this assay depends on their DNA-binding activity and, therefore, dimerization may also be analysed in the presence of DNA. Direct analysis of dimerization is determined by an assay using co-translation of labelled full-length protein and delation mutants or point mutants of the transcription factor (43). The forma-tion of a heterodimer is analysed using immunoprecipitation of labelled protein with a specific antisera which recognizes an epitope on only one of the two proteins. This assay is described in *Protocol 11*.

Protocol 11. Analysis of dimerization by immunoprecipitation

1. Select two different forms of the protein, only one of which contains the epitope for the antibody to be used.
2. Co-translate the protein *in vitro* in the presence of [^{35}S]methionine as described in *Protocol 6*.
3. Prepare Protein A Sepharose beads (PAS) by equilibrating the beads in immunoprecipitation buffer (IB)^a overnight at 4°C. Spin the beads for 10 sec in a microcentrifuge and wash three times with IB.
4. Dilute aliquots of the lysate containing 10^5 c.p.m. in IB in a final volume of 50 μl.
5. Centrifuge the sample for 5 min to pellet aggregates.
6. Add antiserum or pre-immune serum. The amount used depends on the quality of the serum and should be determined empirically. Typically, 1–5 μl of antiserum is used.
7. Incubate at 4°C overnight, or for 30 min at room temperature.
8. Add 50 μl of a 50% slurry of PAS.
9. Incubate at 4°C for 30 min.
10. Pellet the beads by spinning for 10 sec in a microcentrifuge.
11. Wash the beads three times with IB.
12. Add SDS-PAGE sample buffer and analyse by SDS-PAGE.

^{a} 1 × IB: 100 mM KCl, 20 mM Hepes (pH 7.4), 1 mg/ml BSA, 0.1% NP40, 10% glycerol.

The type of experiment described in Protocol 11 is illustrated in *Figure 12*. A [^{35}S]methionine-labelled N-terminal deletion mutant of the mouse oestrogen receptor 121–599 is co-translated with either the labelled full-length protein 1–599 or full-length protein containing specific point mutations. A specific antipeptide antiserum MP15, which has been raised to amino acids 17–28, is then used to immunoprecipitate the truncated form of the protein 121–599, where this deletion mutant is associated with the full-length form in a heterodimer. This approach demonstrates directly that the specific residues leucine-511 and isoleucine-518 form a critical part of the dimer interface in the oestrogen receptor protein.

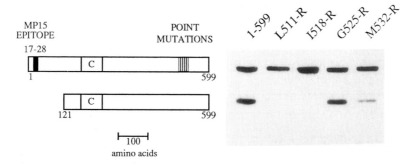

Figure 12. Immunoprecipitation of oestrogen receptor mutant proteins.

6.6 The effect of point mutations on transcriptional activation

A number of classes of activation domains have been identified in different transcription factors (47). These include the acidic activation domains of the yeast activators GAL4 and GCN4, glutamine-rich regions in *Sp*1 and many *Drosophila* homeodomain proteins, proline-rich regions, and also the serine- and theonine-rich activating regions of the POU domain proteins. Of these, the acidic activation regions, which contain relatively high concentrations of acidic amino acids are the best characterized. Mutational analysis indicates that the activation function depends in part on the number of acidic residues and that these regions may form amphipathic α helices (48). The precise organization of this type of domain, however, has still to be determined and it has been proposed that the acidic-rich sequence may exist as an unstructured region with a defined structure only forming on interaction with another regulatory protein (49).

6.6.1 Identification of residues involved in transcriptional activation by point mutagenesis

The criteria described in Section 6.1 have been used in the following example to identify residues involved in transcriptional activation. The C-terminal boundary of sequences involved in hormone-inducible transactivation in the mouse oestrogen receptor has been mapped by deletion analysis to between amino acids 538 and 552. Sequence comparisons of this region with the equivalent regions of other members in the nuclear hormone receptor super-family indicates a conserved sequence which consists of an invariant glutamic acid residue flanked by pairs of hydrophobic residues (39). Comparison with other transactivation domains indicates that the residues in this region could form an amphipathic α helix. Point mutagenesis of these amino acids demonstrates, however, that the hydrophobic residues are essential for hormone-dependent transcriptional activation, while mutation of the conserved acidic residue has no effect in the full-length protein. The structural organization of this region is still to be determined.

7. Summary and conclusions

The methods in this chapter have described the approaches taken to investigate the functional properties of a cloned transcription factor. This includes the initial identification of a cDNA clone by expression in heterologous cells, the mapping of the boundaries of functional domains, and the analysis of individual domains using chimeric proteins and point mutagenesis. The methods have been illustrated with reference to one particular transcription factor, the mouse oestrogen receptor, but the general principles involved are suitable for the analysis of any protein for which a cDNA clone has been isolated.

References

1. Berthois, Y., Katzenellenbogen, J. A., and Katzenellenbogen, B. S. (1986). *Proc. Natl. Acad. Sci. USA*, **83**, 2496.
2. Eckert, R. L. and Katzenellenbogen, B. S. (1982). *Cancer Res.*, **42**, 139.
3. Maniatis, T., Fritsch, E. F., and Sambrook, J. (ed.) (1989). *Molecular Cloning, A. Laboratory Manual*. Cold Spring Harbor Press, Cold Spring Harbor, NY.
4. Morgenstern, J. P. and Land, H. (1990). *Nucleic Acids Res.*, **18**, 1068.
5. Gorman, C. M., Moffet, L. F., and Howard, B. H. (1982). *Mol. Cell. Biol.*, **2**, 1044.
6. Hall, C. V., Jacob, P. E., Ringold, G. M., and Lee, F. (1983). *J. Mol. Appl. Genet.*, **2**, 101.
7. De Wet, J. R., Wood, K. V., Pehuea, M., Helinski, P. R., and Subromani, S. (1987). *Mol. Cell. Biol.*, **7**, 725.
8. Knoll, B. J., Zarucki-Schulz, T., Dean, D. C., and O'Malley, B. W. (1983). *Nucleic Acids Res.*, **11**, 6733.
9. Klein-Hitpass, L., Schorpp, M., Wagner, U., and Ryffel, G. (1986). *Cell*, **46**, 1053.
10. Luckow, B. and Schutz, G. (1987). *Nucleic Acids Res.*, **15**, 5490.
11. Boshart, M., Kluppel, M., Schmidt, A., Schutz, G., and Luckow, B. (1992). *Gene*, **110**, 129.

12. Wigler, M., Silverstein, S., Lee, L.-S., Pellicer, A., Cheng, Y.-C., and Axel, R. (1977). *Cell*, **11**, 223.
13. Chu, G., Hayakawa, H., and Berg, P. (1987). *Nucleic Acids Res.*, **15**, 1311.
14. Felgner, P. L., Gadek, T. R., Holm, M., Roman, R., Chan, H. W., Wenz, M., Northrop, J. P., Ringold, G. M., and Danielson, M. (1987). *Proc. Natl. Acad. Sci. USA*, **84**, 7413.
15. Sleigh, M. J. (1986). *Anal Biochem.*, **156**, 251.
16. Kozak, M. (1987). *Nucleic Acids Res.*, **15**, 8125.
17. Melton, D. A., Krieg, P. A., Rebagliati, M. R., Maniatis, T., Zinn, K., and Green, M. R. (1984). *Nucleic Acids Res.*, **12**, 7035.
18. Summers, M. D. and Smith, G. E. (1987). *A Manual of Methods for Baculovirus Vectors and Insect Cell Procedures.* Texas Agriculture Experiment Station Bulletin, **1555**.
19. Page, M. J. (1989). *Nucleic Acids Res.*, **17**, 454.
20. Fawell, S. E. White, R., Hoare, S., Sydenham, M., Page, M., and Parker, M. G. (1990). *Proc. Natl. Acad. Sci. USA*, **87**, 6883.
21. Kitts, P. A., Ayres, M. D., and Possee, R. D. (1990). *Nucleic Acids Res.*, **18**, 5667.
22. Elliston, J. F., Fawell, S. E., Klein-Hitpass, L., Tsai, S.-Y., Tsai, M.-J., Parker, M. G., and O'Malley, B. O. (1990). *Mol. Cell. Biol.*, **10**, 6607.
23. Mellon, P., Parker, V., Gluzman, Y., and Maniatis, T. (1981). *Cell*, **27**, 279.
24. Krust, A., Green, S., Argos, P., Kumar, V., Walter, P., Bornert, J.-M., and Chambon, P. (1986). *EMBO J.*, **5**, 891.
25. Kumar, V., Green, S., Staub, A., and Chambon, P. (1986). *EMBO J.*, **5**, 2231.
26. Dingwall, C. R. and Laskey, R. A. (1991). *Trends Biochem. Sci.*, **16**, 478.
27. Guiochon-Mantel, A., Lescop, P., Christin-Maitre, S., Loosfelt, H., Perrot-Applanet, M., and Milgrom, E. (1991). *EMBO J.*, **10**, 3851.
28. Guiochon-Mantel, Loosfelt, H. A., Lescop, P., Sar, S., Atger, M., Perrot-Applanet, M., and Milgrom, E. (1989). *Cell*, **57**, 1147.
29. White, R., Lees, J. A., Needham, M., Ham, J., and Parker, M. (1987). *Mol. Endocrinol.*, **1**, 816.
30. Schwabe, J. W. R. and Rhodes, D. (1991). *Trends Biochem. Sci.*, **16**, 291.
31. Lamb, P. and McKnight, S. L. (1991). *Trends Biochem. Sci.*, **16**, 417.
32. Landschulz, W. H., Johnson, P. F., and McKnight, S. L. (1988). *Science*, **230**, 1759.
33. Visvader, J. and Begley, C. G. (1991). *Trends Biochem. Sci.*, **16**, 330.
34. Rauscher III, F. J., Cohen, D. R., Curran, T., Bos, T. J., and Vogt, P. K. (1988). *Science*, **234**, 1010.
35. Fawell, S. E., Lees, J. A., and Parker, M. G. (1989). *Mol. Endocrinol.*, **3**, 1002.
36. Lees, J. A., Fawell, S. E., and Parker, M. G. (1989). *Nucleic Acids Res*, **17**, 5477.
37. Brent, R. and Ptashne, M. (1985). *Cell*, **43**, 729.
38. Cohen, D. R. and Curran, T. (1990). *Oncogene*, **5**, 929.
39. Danielian, P. S., White, R., Lees, J. A., and Parker, M. G. (1992). *EMBO J.*, **11**, 1025.
40. Sadowski, I. and Ptashne, M. (1989). *Nucleic Acids Res.*, **17**, 7539.
41. Lillie, J. W. and Green, M. P. (1989). *Nature*, **338**, 39.
42. Mader, S., Kumar, V., de Verneuil, H., and Chambon, P. (1989). *Nature*, **338**, 39.

43. Fawell, S. E., Lees, J. A., White, R., and Parker, M. G. (1990). *Cell*, **60**, 953.
44. Mangelsdorf, D. J., Ong, S. E., Dyck, J. A., and Evans, R. M. (1987). *Nature*, **345**, 224.
45. McPherson, M. J. (ed.) (1991). *Directed Mutagenesis: A Practical Approach*. IRL Press, Oxford.
46. Nakamaye, K. and Eckstein, F. (1986). *Nucleic Acids Res.*, **14**, 9679.
47. Herr, W. (1991). In *Molecular Aspects of Cellular Regulation*, Vol. 6 (ed. P. Cohen and J. G. Foulkes), pp. 25–52. Elsevier Science Publishers, Amsterdam.
48. Giniger, E. and Ptashne, M. (1987). *Nature*, **330**, 670.
49. Frankel, A. D. and Kim, P. S. (1991). *Cell*, **65**, 717.

A1

Physical characterization of DNA-binding proteins in crude preparations

MIN LI and STEPHEN DESIDERIO

1. Introduction

After identification of a DNA-binding protein and characterization of its binding specificity, the investigator's next goal is usually to obtain pure protein for functional studies and for amino acid sequence analysis. Before this is achieved, it is often desirable to determine the size of the binding protein. It is also possible, in relatively crude preparations, to determine whether the protein binds DNA as a monomer or as a multimer. Knowledge of these simple physical characteristics can affect the choice of methodologies for purification of the protein and for molecular cloning of the gene that encodes it.

2. Determination of molecular weight from hydrodynamic measurements

2.1 Theoretical and practical considerations

Molecular weight is related to three other properties of a macromolecule— the sedimentation coefficient, diffusion coefficient, and partial specific volume—by the Svedberg equation:

$$M = \frac{RTs}{D(1 - \bar{v}\rho)} \qquad (1)$$

where M is the molecular weight, s is the sedimentation coefficient, D is the diffusion coefficient, $\bar{v}$ is the partial specific volume, and ρ is the solvent density. It follows that if s, D and $\bar{v}$ can be determined, the molecular weight of a macromolecule can be calculated. For proteins in crude preparations, s and D are readily determined by velocity sedimentation and gel filtration chromatography, respectively; the partial specific volume, however, is generally inaccessible under these conditions. Fortunately, for most proteins this value varies within a narrow range, and the molecular weight of an impure protein can, therefore, often be estimated from experimentally determined values of s and D and an assumed partial specific volume ($0.725\,g/cm^3$) (1).

The hydrodynamic parameters needed to estimate the molecular weight of a specific DNA-binding protein can be determined even in very crude preparations. The Stokes' radius (a) is related to the diffusion coefficient (D) by the Stokes–Einstein equation,

$$D = \frac{KT}{6\pi\eta a} \tag{2}$$

where K is Boltzman's constant, T is the absolute temperature, and η is the viscosity of the buffer at T. As described below, the Stokes' radius of an unknown protein is readily estimated from its gel filtration partition co-efficient, K_{av}, by comparison with a standard curve; from the Stokes' radius, a diffusion coefficient is calculated. The sedimentation coefficient (s) is likewise determined by comparing the protein's mobility in a centrifugal field with the mobilities of standard proteins of known s value.

Because these measurements are generally made under non-denaturing conditions, the molecular weight of the native protein is obtained. Thus, in combination with denaturing methods (see below), one can infer whether the DNA-binding protein exists as a stable multimer. One pitfall of this approach is the potential for non-specific interactions with other proteins in the crude preparation, which can affect the mobility of the DNA-binding protein during gel filtration and velocity sedimentation. In the protocols given below, we therefore suggest that these procedures be performed at salt concentrations of 100 mM or greater.

2.2 Estimation of Stokes' radius by gel filtration

2.2.1 Partial purification of the DNA-binding protein

It is useful to effect a partial purification of a DNA-binding protein before proceeding to characterization of its hydrodynamic properties because the resulting increase in specific activity makes it easier to monitor specific DNA binding in a relatively small amount of protein. We favour chromatography on BioRex 70 or other cation exchange matrices for this purpose, because of their high capacity and because nucleic acids are generally not retained on cation exchangers at neutral pH. Thus, a DNA-binding protein that is retained on BioRex 70 is usually separated from any nucleic acids present in the crude extract, which might otherwise interfere with assays for specific binding.

Protocol 1. Partial purification of the DNA-binding protein by BioRex 70 chromatography

1. Prepare a nuclear extract from a source rich in specific DNA-binding activity. The extraction protocol will, of course, depend on the protein to be characterized; we have elsewhere described a detailed method for preparation of nuclear extracts from calf thymus (2). Determine the concentration of protein in the extract; we routinely use the assay of Bradford (3). A protease inhibitor (such as 0.2 mM phenylmethylsulphonyl fluoride (PMSF), added fresh from a 1 M stock in ethanol) should be present during the extraction, BioRex 70 chromatography and the subsequent analytic procedures described below.

2. BioRex 70 (200–400 mesh) is obtained from BioRad. Rinse the BioRex 70 matrix briefly with 0.5 M NaOH and then wash repeatedly with a buffer containing 50 mM NaCl, 50 mM Hepes–NaOH (pH 7.5), 10 mM 2-mercaptoethanol, and 2 mM EDTA, until the pH of the effluent reaches 7.5.

3. For an analytic trial, pour a 5 ml column (approximately 1 cm × 6 cm) and equilibrate with at least 10 bed volumes of buffer B (50 mM NaCl, 20 mM Hepes–NaOH (pH 7.5), 10 mM 2-mercaptoethanol, 2 mM EDTA, 0.2 mM PMSF and 10% (w/v) glycerol). The capacity of BioRex 70 is 2–5 mg crude protein for every 1 ml of bed volume.

4. Load 10 mg nuclear protein (concentration 5–15 mg/ml) in buffer B at a flow rate of 10 ml/h. Collect 0.5 ml fractions. Wash the column with 5 ml buffer B. Develop the column with a 25 ml, linear gradient of NaCl from 50 mM to 1 M in buffer B. Monitor the protein in each fraction by UV absorbance at 280 nm. Determine the NaCl concentration in each fraction by measuring conductivity and comparing the measurements to a standard curve.

5. Identify fractions containing the specific DNA-binding activity by electrophoretic mobility shift, footprinting, or filter-binding assay as appropriate (see Chapters 1 and 2). Dialysis may be necessary before assaying activity, depending on the sensitivity of the binding reaction to salt and the NaCl concentration at which the binding activity is eluted. A concentrating step is usually not required.

Assuming the DNA-binding activity is retained on the column and elutable by salt a refined chromatographic protocol can be developed to optimize purification, once the NaCl concentration at which the activity elutes has been determined. We have used the following sequence, after the sample is loaded: (a) an NaCl step gradient from 50 mM to a concentration 50 mM below the elution point; (b) a shallow, 100 mM linear gradient from 50 mM below the elution point to 50 mM above it; (c) a step gradient to 1 M NaCl, which regenerates the column. The use of a shallow gradient around the elution point enhances separation of the binding protein from other proteins with similar chromatographic properties.

DNA-binding activity that is not retained on the column at pH 7.5 may be retained at a lower pH. Alternatively, a DEAE matrix can be used, but because nucleic acids are retained on DEAE at low salt concentration, two problems may arise: (a) DNA or RNA bound to the DEAE matrix can have unreproducible effects on the elution of DNA-binding proteins and (b) nucleic acids may contaminate eluted DNA-binding proteins and interfere with binding assays.

2.2.2 Gel filtration chromatography

The elution position of a protein during gel filtration chromatography is most closely correlated with its Stokes' radius (*a*) (4). The cross-linked agarose matrix Superose (Pharmacia-LKB) provides highly reproducible separations. Two porosities are available: Superose 6 (separation range about 5×10^3–5×10^6 mol. wt.) and Superose 12 (separation range about 1×10^3–3×10^5 mol. wt.). These are commonly obtained in pre-packed columns for use with an FPLC system (Pharmacia-LKB) equipped with a UV monitor and fraction collector.

Protocol 2. Gel. filtration chromatography on Superose 12

1. A pre-packed Superose 12 HR10/30 column (10 mm × 300 mm) is equilibrated with a buffer containing NaCl at a concentration of at least 100 mM. The flow rate should be about 0.3 ml/min, to prevent high back pressures that can distort the packed matrix. A trial run should be performed before the protein sample is loaded. The UV trace should be examined for any deflections from baseline that are not dependent on the presence of protein; such deflections can result from transient pressure changes. In most instances there should be only one such deflection, at about 0.2 ml, which is caused by a pressure change resulting from valve switching after the sample is loaded.

2. Load an aliquot (≤150 µl) of the partially purified protein preparation through a 200 µl injection loop. Brief centrifugation may be necessary to remove any insoluble material before chromatography.

3. Develop the column with a 25 ml isocratic gradient containing NaCl at 100 mM or greater, to minimize the possibility of non-specific protein interactions, at a flow rate of 0.2–0.5 ml/min. Collect 0.5 ml fractions.

4. Assay DNA-binding activity in each fraction that elutes after 6 ml (the void volume of the column is about 7 ml).

5. Under conditions identical to those used for the DNA-binding protein, perform an individual chromatographic run for each of a series of standard proteins, loading 30–100 µg in each case. The following standards are commonly used: ferritin ($a = 61.0 \times 10^{-8}$ cm), catalase ($a = 52.2 \times 10^{-8}$ cm), bovine serum albumin (BSA) ($a = 35.5 \times 10^{-8}$ cm), ovalbumin ($a = 30.5 \times 10^{-8}$ cm), and chymotrypsinogen ($a = 20.9 \times 10^{-8}$ cm). After

performing the individual runs, combine the standards and chromatograph under the same conditions.

6. Determine the void volume, V_0, by chromatographing blue dextran (detected by absorbance at 280 nm) under the same conditions as above. The elution volume of blue dextran is equivalent to V_0. The total volume of pre-packed Superose columns, V_t, is provided by the supplier.

7. Calculate the partition coefficient (K_{av}) for each standard and for the DNA-binding activity, using the following formula:

$$K_{av} = (V_e - V_0)/(V_t - V_0) \tag{3}$$

where V_0 is the void volume of the column, V_t is the total volume of the column, and V_e is the elution volume of each species. Plot a standard curve relating K_{av} to a; this should be linear under the conditions described above. From the slope of this curve and the observed K_{av} of the DNA-binding activity, a Stokes' radius can be calculated.

2.3 Estimation of Svedberg constant by velocity sedimentation

A protein's sedimentation coefficient can be measured directly by determining its velocity in a centrifugal field, but for an impure protein in limiting supply this value is more easily estimated by comparing its mobility to the mobilities of a set of proteins of known s value. In generating the standard curve, $s_{20,w}$ values (i.e. sedimentation coefficients extrapolated to water at 20°C) are generally used. Such a curve, therefore, provides the $s_{20,w}$ value for the protein under study. Subsequent calculation of molecular weight is facilitated when the s value is in this form because the viscosity and density of water at 20°C are defined.

Protocol 3. Glycerol gradient sedimentation

1. Form two identical, 15–30% linear glycerol gradients (vol. 4.9 ml) in 5 ml centrifuge tubes for the Beckman SW50.1 rotor. The buffer should contain NaCl at 100 mM or greater, to minimize non-specific protein interactions.

2. Layer 100 μl of the partially purified protein preparation on top of one gradient. In determining the minimum amount of sample required, assume that activity will be diluted about 10-fold after centrifugation.

3. On the other gradient, layer 100 μl of a solution containing 20 μg of each of the following standard proteins: catalase ($s_{20,w} = 11.3 \times 10^{-13}$ sec), aldolase $s_{20,w} = 8.3 \times 10^{-13}$ sec), BSA ($s_{20,w} = 4.2 \times 10^{-13}$ sec), ovalbumin ($s_{20,w} = 3.6 \times 10^{-13}$ sec), and chymotrypsinogen ($s_{20,w} = 2.5 \times 10^{-13}$ sec).

4. Centrifuge gradients in the Beckman SW50.1 rotor at 45 000 r.p.m. for

Protocol 3. *Continued*

26 h at 4°C. Collect 100 μl fractions. If an automated gradient collector is not available, a micropipette can be used to remove fractions from top to bottom. Alternatively, puncture the bottom of the tube with a 19 gauge needle and collect 5–10 drops per fraction.

5. Localize the DNA-binding activity by standard binding assays. To determine elution positions of the protein standards, fractionate a 15 μl aliquot of each fraction by SDS-polyacrylamide gel electrophoresis (SDS-PAGE) and visualize protein by staining with Coomassie blue. Generate a standard curve by plotting the known sedimentation coefficients ($s_{20,w}$) against volume or fraction number. For the standard proteins and conditions described here, a linear relationship between $s_{20,w}$ and volume should be observed. From the slope of the standard curve and the observed sedimentation position, the $s_{20,w}$ of the DNA-binding protein is determined.

2.4 Calculation of molecular weight from Stokes' radius and Svedberg constant

By combining Equations 1 and 2, molecular weight is related to Stokes' radius, sedimentation coefficient, and partial specific volume:

$$M = \frac{6 \pi \eta Nas}{(1 - \bar{v}\rho)} \tag{4}$$

(where $N = 6.02 \times 10^{23}$). For $s = s_{20,w}$, $\eta = 0.010019$ g/sec·cm (the viscosity of water at 20°C), and $\rho = 0.998$ g/ml (the density of water at 20°C). In most instances, $\bar{v}$ will be undefined, and an assumed value must be used. By using a value of 0.725 cm³/g, the molecular weights of most proteins can be estimated fairly accurately, but a large deviation from this value can lead to a result that is grossly inaccurate; molecular weights determined by this method should, therefore, be regarded as tentative.

From the molecular weight (M) and Stokes' radius (a) obtained by means described above, a frictional ratio for the DNA-binding protein, f/f_0, can be estimated from the equation:

$$f/f_0 = \frac{a}{(3\bar{v}M/4\pi N)^{1/3}} \tag{5}$$

The frictional ratio relates the frictional coefficient of the molecule under study to the frictional coefficient of a sphere of similar mass and partial specific volume. The structural interpretation of this ratio is complicated because the frictional coefficient is not only determined by the shape of a protein but also by its degree of hydration. The frictional ratio is of practical importance, however, because an aberrantly small ratio (i.e. a value <1) should alert the investigator to inaccuracies in the determination of Stokes' radius, sedimentation coefficient, or both. For example, adsorption of the DNA-binding protein to the gel filtration matrix will lead to an underestimate of the Stokes' radius and an aberrantly small frictional ratio (1).

To illustrate the overall procedure, consider the following example. A Superose 12 HR10/30 column ($V_t = 20.3$ ml) is determined to have a V_0 of 7.1 ml. A DNA-binding protein elutes from this column at a volume of 13.2 ml. From Equation 3, $K_{av} = (13.2 - 7.1)/(20.3 - 7.1) = 0.46$. After determining experimentally the V_e for each of several standard proteins of known Stokes' radius, values of K_{av} are calculated and plotted as a function of Stokes' radius. On this curve, let us suppose that a K_{av} of 0.46 corresponds to a Stokes' radius of 31.5×10^{-8} cm. For the same DNA-binding protein, $s_{20,w}$ is estimated to be 4×10^{-13} sec, by glycerol gradient centrifugation and comparison to a standard curve, as described in *Protocol 3*. Assuming a partial specific volume of 0.725 cm^3/g, the native molecular weight (M) is estimated using Equation 4:

$$M = \frac{6\pi(0.01 \text{ g/sec·cm})(6.02 \times 10^{23} \text{ atoms/g·atom})(31.5 \times 10^{-8} \text{ cm}) \, (4 \times 10^{-13} \text{ sec})}{1 - (0.725 \text{ cm}^3/\text{g})(1.0 \text{ g/cm}^3)}$$

$$= 52\,000$$

The frictional ratio of the protein, f/f_0, is then estimated by Equation 5 to be:

$$\frac{31.5 \times 10^{-8} \text{ cm}}{\left[\frac{(3)(0.725 \text{ cm}^3/\text{g})(5.2 \times 10^4)}{4\pi(6.02 \times 10^{23} \text{atoms/g·atom})} \right]^{1/3}} = 1.28$$

3. Sizing of the DNA-binding component by SDS-polyacrylamide gel electrophoresis

3.1 Recovery of DNA-binding activity by protein renaturation after SDS-polyacrylamide gel electrophoresis

3.1.1 Theoretical and practical considerations

In many instances, the enzymatic activity (5) or specific ligand-binding activity (6) of a denatured protein can be recovered, at least partially, by renaturation *in vitro*. Such is the case for many DNA-binding proteins (2, 7, 8). If specific binding is renaturable and is accomplished by a single polypeptide species (either as a monomer or a homomeric multimer), the molecular weight of the polypeptide responsible for binding can be estimated by SDS-PAGE. This approach will obviously not provide the size of the native protein, and is subject to the usual reservations regarding molecular weight values obtained by SDS-PAGE (for example, that post-translational modifications such as glycosylation and phosphorylation can affect mobility). None the less, the information obtained by this procedure can significantly affect the subsequent choice of molecular cloning methodology. For example, a DNA-binding protein whose activity can be recovered in a single fraction after SDS-PAGE is likely to require only a single polypeptide species for binding. This information is of practical importance because one common means of identifying cDNAs that encode specific DNA-binding proteins involves expression cloning in recombinant bacteriophage (see Chapter 5). For this method to succeed, it is necessary that binding be accomplished by a single polypeptide species (i.e. the product of a single gene).

3.1.2 Elution and recovery of DNA-binding activity after SDS-PAGE

Protocol 4. Electrophoresis, elution, and renaturation of DNA-binding protein

1. Denature a sample of crude or partially purified extract containing the DNA-binding activity by heating to 100°C for 2 min in a buffer containing 100 mM 2-mercaptoethanol, 60 mM Tris–HCl (pH 6.8), 10% glycerol, 1% SDS, and 0.0005% bromophenol blue. The amount of sample used should be at least 100 times that required to detect specific DNA-binding activity in a standard assay.

2. Fractionate the reduced, denatured protein and molecular weight markers by electrophoresis through an SDS-polyacrylamide slab gel of thickness 1.5 mm, using standard protocols (9). After electrophoresis, cut off the lane containing molecular weight standards, measure the length of the resolving gel carefully, and stain with Coomassie blue. Alternatively, pre-stained marker proteins (Sigma) may be used, facilitating their alignment with the DNA-binding protein, but in this instance mobility differences arising from pre-staining must be taken into account. These differences in apparent molecular weight are documented by the supplier.

3. Cut the lane containing the DNA-binding protein into 5 mm slices from top to bottom.

4. Transfer each slice to a 0.5 ml microcentrifuge tube that has been punctured at the bottom with a 20 gauge needle. Place each 0.5 ml tube into a 1.7 ml microcentrifuge tube and centrifuge for 5 min at 12 000 r.p.m. This will macerate slices by forcing them through the puncture holes and into the collection tubes. Incubate the gel samples in 2 volumes of a buffer containing 150 mM NaCl, 20 mM Hepes–NaOH (pH 7.5). 5 mM dithiothreitol (DTT), 0.1 mM EDTA, 0.1 mg/ml BSA, and 0.1% SDS for 3 h at room temperature or 1 h at 40°C. Centrifuge at $10\,000 \times g$ for 15 min at room temperature and save the supernatants, which contain eluted protein. Centrifuge the supernatants once again to remove any residual polyacrylamide particles.

5. Add 4 volumes of cold ($-20°C$) acetone to the supernatants and incubate in a dry-ice–ethanol bath for 45 min. Recover the eluted proteins by centrifugation at $12\,000 \times g$ for 15 min. Wash pellets with a 4:1 mixture of acetone and a buffer containing 150 mM NaCl, 20 mM Hepes–NaOH (pH 7.5), 5 mM DTT, 0.1 mM EDTA, and 0.1 mg/ml BSA. This will remove residual SDS. Dry the pellets in air.

6. Dissolve the pellets in 5 µl of a buffer containing 6 M guanidine–HCl, 150 mM NaCl, 20 mM Hepes–NaOH (pH 7.5), 5 mM DTT, 0.1 mM EDTA, and 0.1 mg/ml BSA. Incubate at room temperature for 20 min. Dilute samples 50-fold with the same buffer lacking guanidine–HCl and incubate for an additional 12 h at room temperature or 4°C. Specific DNA-binding activity can be assayed directly.

7. The apparent molecular weight of the polypeptide responsible for DNA binding can be estimated by comparing the position of the active gel slice(s) to the mobilities of the molecular weight standards (see *Figure 1*). If the standards were detected by staining after electrophoresis, mobility measurements must be corrected for any change in gel size that might have occurred during staining and destaining.

The renaturation procedure described in this protocol is based on a method developed by Hager and Burgess (5). Efficiency of renaturation is affected by temperature; therefore, for an uncharacterized protein we suggest that the initial renaturation experiments be performed both at room temperature and at 4°C. Especially when working with crude extracts, it is advisable to demonstrate that the binding activity eluted from the SDS-polyacrylamide gel has the same specificity as the one under study (see *Figure 1*).

3.1.3 Detection of the specific DNA-binding species after SDS-PAGE and transfer to nitrocellulose

After separation by SDS-PAGE and transfer to nitrocellulose, a large subset of DNA-binding proteins retain their ability to bind DNA. On this basis, several methods for direct detection of sequence-specific DNA-binding proteins have been developed (10–12). These methods have the following common features:

(a) electrophoretic separation of proteins by SDS-PAGE

(b) transfer of protein to a nitrocellulose filter

(c) a blocking step to prevent non-specific adsorption of DNA to nitro-cellulose and protein

(d) incubation of the blocked filter with a radiolabelled probe containing the recognition sequence

(e) removal of unbound DNA and detection of bound DNA by autoradi-ography

Appendix 1

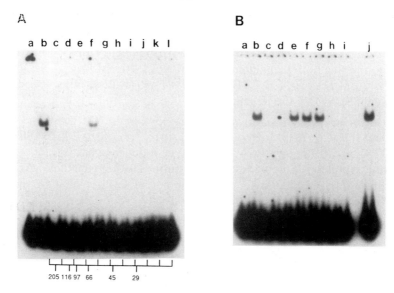

Figure 1. Renaturation of a specific DNA-binding activity after SDS-PAGE. The protein recognizes the nonamer recombinational signal sequence of immunoglobulin gene segments. (A) Partially purified protein (600 μg) was fractionated by electrophoresis through SDS-polyacrylamide. The lane containing the protein was sliced, protein was eluted and renatured as described in *Protocol 4*. Renatured protein was assayed for specific binding to a nonamer-containing probe by electrophoretic mobility shift assay. Lane a, no protein; lane b, 2 μg of the loaded sample; lanes c–l, 5 μl of protein from each gel slice. The positions of molecular weight standards in relation to the gel slices assayed are indicated at the bottom of the figure. (B) Renatured binding activity retains specificity. The active gel fraction was assayed for binding to a nonamer-containing probe (10 pg/reaction) in the presence of specific competitor fragments. Lane a, no protein or specific competitor added; lane b, no specific competitor; lanes c–i, assays performed in the presence of 1500 pg of wild-type competitor (lane c), competitors with mutations outside the nonamer site (lanes d, h, i), or competitors with altered nonamer sequences (lanes e, f, g). Lane j, 2 μg of the loaded sample, assayed in the absence of specific competitor. (Reproduced with permission from reference 2.)

The advantages of these methods are their ease and their high resolving power. Two drawbacks are a relatively high background and a preference for long-lived protein–DNA complexes.

Refinements to the approach as originally described (10) have included electrophoretic transfer of protein to nitrocellulose, the use of non-fat dry milk as a blocking agent, inclusion of non-specific competitor DNA in the binding reaction, and enhancement of filter-bound DNA-binding activity by a cycle of protein denaturation and renaturation (11, 12). The following protocol incorporates these refinements.

Protocol 5. Detection of specific DNA-binding proteins after electrophoresis and transfer to nitrocellulose

1. Heat the protein sample to 100°C for 5 min in SDS-PAGE sample buffer (100 mM 2-mercaptoethanol, 60 mM Tris–HCl (pH 6.8), 10% glycerol, 1% SDS, 0.0005% bromophenol blue). Fractionate by electrophoresis through a 10% SDS-polyacrylamide gel.

2. Soak the gel for 30 min in 25 mM Tris, 190 mM glycine (pH 8.3), 20% (v/v) methanol. Transfer protein electrophoretically to a nitrocellulose filter in the same buffer at 250 mA for 2 h at 4°C.

3. Block the filter by bathing for 1 h in a solution containing 5% non-fat dry milk (Carnation), 50 mM Tris–HCl (pH 7.5), 50 mM NaCl, 1 mM EDTA, 1 mM DTT.

4. Wash the filter twice (5 min each wash) with 50 mM NaCl, 10 mM Tris–HCl (pH 7.5), 1 mM EDTA, 1 mM DTT.

5. Denature the protein bound to the filter by incubating in a solution containing 6 M guanidine–HCl, 50 mM Tris–HCl (pH 8.3), 50 mM DTT, 2 mM EDTA, 0.1% Nonidet P-40 (NP40), 0.25% (w/v) non-fat dry milk for 1 h at 25°C.

6. Allow the proteins to renature in 50 mM Tris–HCl (pH 7.5), 100 mM NaCl, 2 mM DTT, 2 mM EDTA, 0.1% NP40, 0.25% non-fat dry milk for 16 h at 4°C.

7. Rinse the filter in a buffer appropriate for binding of the DNA-binding protein under study. Incubate the filter for 60 min at room temperature in binding buffer containing ^{32}P-labelled probe (specific activity about 1×10^6 c.p.m./ml; binding site concentration about 0.1 nM) and 10 μg/ml poly(dI-dC)·(dI-dC).

8. Wash the filter for a total of 30 min at room temperature with four changes of binding buffer.

9. Detect filter-associated DNA by autoradiography.

The level of non-specific adsorption depends in part on how the probe is prepared. Nick-translated probes generally give higher backgrounds because of the presence of radiolabelled single-stranded DNA. This problem can be minimized by using gel-purified probes that have been labelled by poly-nucleotide kinase or by fill-in synthesis with *E. coli* DNA polymerase I large fragment (Klenow fragment). Even so, non-specific binding of radiolabelled probe is often observed, despite the presence of a large excess of poly(dI-dC)·(dI-dC). To distinguish between specific and non-specific interactions, one should demonstrate that unlabelled, functional binding sites compete with the labelled probe for binding. This may be accomplished by control experiments in which an excess of unlabelled wild-type or mutant sites are present in the binding reactions.

In the protocol given above, filter-bound protein is denatured in guanidine-HCl and allowed to renature before the binding reaction is carried out. While this procedure has been observed to enhance binding by some proteins, the binding activity of other proteins may not survive such treatment. Thus, if a negative result is obtained with the guanidine-HCl method, the procedure should be repeated, omitting steps 5 and 6.

3.2 UV cross-linking

3.2.1 Theoretical and practical considerations

Irradiation of DNA substituted with 5-bromodeoxyuridine (BrdU) generates reactive uracilyl radicals by debromination; these radicals can form adducts with a large number of amino acids, including cysteine, phenylalanine, tyrosine, histidine, lysine, and arginine (13). Thus, UV irradiation of a complex between a DNA-binding protein and BrdU-substituted DNA can result in formation of a covalent DNA–protein adduct (14). Photochemical attachment to DNA can be exploited to determine the apparent molecular weight of a DNA-binding polypeptide, using the following general scheme:

(a) synthesis of BrdU-substituted, ^{32}P-labelled DNA containing the recognition site

(b) formation of a specific complex between the binding protein and the labelled, substituted DNA fragment

(c) irradiation with UV light

(d) removal of most of the DNA by nuclease treatment

(e) fractionation of products by SDS-PAGE

(f) detection of protein–oligonucleotide adducts by autoradiography

This approach has some advantages over elution–renaturation methods: the DNA-binding activity need not be able to survive denaturation and renaturation, and fewer manipulations are required. An advantage of cross-linking methods over nitrocellulose filter binding is that complexes with relatively short half-lives can still be detected. Both the elution–renaturation and filter transfer methods require that binding be accomplished by a single polypeptide or, if more than one polypeptide is required, that these have identical mobilities on SDS-PAGE. The UV cross-linking method does not make this demand, since binding is accomplished before denaturation and fractionation of the active species. The cross-linking method requires the presence of T residues within or close to the recognition sequence. A potential drawback is the possibility of spurious cross-linking, but unambiguous results can be obtained even with relatively crude preparations of DNA-binding protein, provided that sufficient non-specific competitor DNA is included in the binding reaction and that appropriate controls for binding specificity are included (8).

3.2.2 Synthesis of BrdU-substituted, ^{32}P-labelled probes

Although unsubstituted DNA can undergo photochemical attachment to proteins, BrdU-substituted DNA is far more reactive. Therefore, the first step in photochemical cross-linking experiments is usually synthesis of BrdU-substituted, radiolabelled target DNA. In the protocol given below, a short oligonucleotide is used to prime synthesis of substituted DNA on a longer synthetic template that spans the recognition site for the DNA-binding protein. An advantage of this approach is that it is not necessary to treat the photoadduct with nucleases before fractionation by SDS-PAGE.

Protocol 6. Synthesis of BrdU-substituted, ^{32}P-labelled oligonucleotide probe

1. Synthesize a 30- to 35-mer template oligonucleotide spanning the DNA-binding site, and a second, primer oligonucleotide (12-mer) complementary to the 3′-terminal 12 nucleotides of the template.

2. Combine 10 μg of template oligonucleotide and 5 μg primer oligonucleotide in a 100 μl reaction containing 50 mM Tris–HCl (pH 7.5), 10 mM $MgCl_2$, 1 mM DTT, and 50 μg/ml BSA. Anneal by heating to 100°C for 2 min, followed by sequential incubation at 65°C for 10 min, 37°C for 10 min, and room temperature for 10 min.

3. To 10 μl of the annealing reaction, add 2 μl each 3 mM dATP, dGTP, and 5-bromo-2′-deoxyuridine triphosphate. Add 80 μl Klenow buffer (50 mM Tris–HCl (pH 7.2), 10 mM $MgSO_4$, 0.1 mM DTT, 50 μg/ml BSA), 3 μl [α-^{32}P]dCTP (10 mCi/ml; 3000 Ci/mmol), and 2 μl *E. coli* DNA polymerase I Klenow fragment (5 units/μl). Incubate at 23°C for 30 min. Bring the reaction volume to 100 μl with TE buffer (pH 7.5) (10 mM Tris–HCl (pH 7.5), 1 mM EDTA) and extract once with buffered phenol. Precipitate the DNA in ethanol and fractionate by electrophoresis through a 12% polyacrylamide gel.

4. Locate the ^{32}P-labelled, BrdU-substituted fragment by autoradiography, excise, and electro-elute from the gel slice. To electro-elute the probe, place the gel slice into a dialysis bag (0.25 inch diameter) with 0.5 ml TBE buffer (90 mM Tris–borate (pH 7.5), 2 mM EDTA). Immerse the bag in a horizontal gel electrophoresis apparatus containing TBE, with the long axis of the bag perpendicular to the electric field. Elute at a voltage gradient of about 10 V/cm for 30 min. Reverse the direction of the field for 30 sec and remove the buffer, which contains the eluted fragment.

5. Collect the electro-eluted DNA by precipitation in ethanol and dissolve the DNA pellet in 50 μl TE (pH 7.5).

In the final precipitation step, an inert carrier such as linear polyacrylamide (LPA), which does not interfere with DNA-protein binding, should be added (2 μl of a 5 mg/ml LPA solution is sufficient for precipitation from volumes of up to 1 ml).

As an alternative to the use of specific oligonucleotides, the following general scheme may be used. A DNA fragment containing the binding site is cloned into a filamentous bacteriophage-based vector such as pBluescript (Stratagene) and single-stranded DNA is rescued by transformation of an F⁺ *E. coli* host and infection with helper bacteriophage. Protocols for rescue of single-stranded DNA from such vectors are provided by the supplier. A probe is then synthesized on the single-stranded template using an appropriate oligonucleotide primer (for example, the T3 or T7 sequencing primers of pBluescript) and *E. coli* DNA polymerase I large fragment in presence of 50 μM each dATP, dGTP, 5-bromo-2'-deoxyuridine triphosphate and 5 μM $[\alpha$-$^{32}P]$dCTP. The probe is then released from the vector by cleavage with the appropriate restriction enzyme(s) and purified by polyacrylamide gel electrophoresis.

With either approach, before proceeding to UV cross-linking the substituted probe should be tested for binding to the protein under study. The relative affinities of the substituted and unsubstituted probes can be assessed by comparing the ability of unlabelled, unsubstituted DNA to compete for binding to either radiolabelled probe.

3.2.3 Sizing of a specific DNA-binding protein by UV cross-linking

---

Protocol 7. Formation of specific protein–oligonucleotide adducts and sizing by SDS-PAGE

1. Combine 300–500 μg active protein fraction from BioRex 70, 15 ng radiolabelled, substituted 30-mer probe (4×10^6 Cerenkov c.p.m.) and 10 μg poly(dI-dC)·(dI-dC) in a 400 μl reaction volume under buffer conditions suitable for the binding protein under study. Allow binding to proceed to equilibrium (typically 30 min at room temperature or 30°C).

2. Layer a parafilm sheet over a bed of crushed ice and transfer 100 μl aliquots of the binding reaction on to the sheet. Irradiate aliquots for 0, 15, 30, or 60 min with a Fotodyne UV lamp (maximum emission 310 mm) at a distance of 1 cm.

3. Add deoxycholate to 0.5 mg/ml and trichloroacetic acid to 12%. Incubate on ice for 15 min and collect precipitates by centrifugation at 4°C in a microcentrifuge.

4. Dissolve precipitates in 20 μl sample buffer (100 mM 2-mercaptoethanol, 60 mM Tris–HCl (pH 6.8), 10% glycerol, 1% SDS, 0.0005% bromophenol blue) and add 2 μl unadjusted Tris base (pH about 9.0) to neutralize.

Boil samples for 3 min and fractionate by SDS-PAGE. Include molecular weight standards.

5. Stain molecular weight markers with Coomassie blue, dry the gel, and detect radiolabelled protein(s) by autoradiography.

Covalent linkage to an oligonucleotide 35 nucleotides long, or less, has little effect on the mobility of proteins in SDS-polyacrylamide gels (15), and the use of an oligonucleotide probe obviates the need for nuclease digestion before electrophoresis. When a larger probe is used, nuclease treatment of protein–DNA adducts is required. For example, the following procedure was used by Sharp and co-workers to determine the size of the adenovirus major late transcription factor (MLTF) in an impure state (8). A crude preparation of binding protein (60 μg total protein) was incubated with 1 ng radiolabelled, BrdU-substituted DNA fragment (about 300 bp), and 5 μg poly(dI-dC)·(dI-dC) in a 50 μl reaction volume in MLTF binding buffer (12 mM Hepes–NaOH (pH 7.9), 12% glycerol, 60 mM KCl, 5 mM $MgCl_2$ 4 mM Tris–HCl, 0.6 mM EDTA, 0.6 mM DTT). After irradiating with UV light essentially as described above, $CaCl_2$ was added to 10 mM. Following addition of 3.3 μg DNase I (Worthington) and 1 IU micrococcal nuclease (Worthington), digestion was allowed to proceed for 30 min at 37°C before fractionation of products by SDS-PAGE.

In crude extracts, especially at prolonged irradiation times, multiple polypeptide–DNA adducts are sometimes observed. To differentiate between non-specific and specific protein–DNA interactions, control binding reactions should be run in the presence of excess unlabelled probe containing a wild type or mutant (non-functional) binding site. Formation of specific DNA–protein adducts should be inhibited by excess wild-type probe, but not by the mutant probe.

3.3 Comparison of hydrodynamic and gel electrophoretic measurements

In many instances, the molecular weight of the native protein will agree with that determined by one of the SDS-PAGE methods described above. In such cases, the protein is likely to be a monomer, at least under the conditions in which the native molecular weight determination was made. This qualification is important because the stability of multimeric protein complexes is salt-dependent. Thus, hydrodynamic measurements made at an ionic strength of 0.1 M may in some instances yield a much larger estimate of molecular weight than similar measurements made at 0.5 M salt. With crude preparations, however, it is difficult to assess increases in apparent, native molecular size at lower ionic strengths because of the possibility of non-specific protein–protein interactions. Thus, purification of the DNA-binding protein is required before its native molecular weight can be determined definitively and compared with the denatured molecular weight of the polypeptide responsible for binding (see Chapter 4).

References

1. Siegel, L. M. and Monty, K. J. (1966). *Biochem. Biophys. Acta*, **112**, 346.
2. Li, M., Morzycka-Wroblewska, E., and Desiderio, S. V. (1989). *Genes Dev.*, **3**, 1801.
3. Bradford, M. (1976). *Anal. Biochem.*, **73**, 248.
4. Ackers, G. K. (1964). *Biochemistry*, **3**, 723.
5. Hager, D. A. and Burgess, R. R. (1980). *Anal. Biochem.*, **109**, 76.
6. Copeland, B. R., Richter, R. J., and Furlong, C. E. (1982). *J. Biol. Chem.*, **257**, 15065.
7. Briggs, M. R., Kadonaga, J. T., Bell, S. P., and Tjian, R. (1986). *Science*, **234**, 47.
8. Chodosh, L. A., Carthew, R. W., and Sharp, P. A. (1986). *Mol. Cell. Biol.*, **6**, 4723.
9. Harlow, E. and Lane, D. (1988). *Antibodies: A Laboratory Manual.* Cold Spring Harbor Laboratory, Cold Spring Harbor, NY.
10. Bowen, B., Steinberg, J., Laemmli, U. K., and Weintraub, H. (1980). *Nucleic Acids Res.*, **8**, 1.
11. Miskimins, W. K., Roberts, M. P., McClelland, A., and Ruddle, F. H. (1985). *Proc. Natl. Acad. Sci. USA*, **82**, 6741.
12. Singh, H., LeBowitz, J. H., Baldwin, A. S., Jr., and Sharp, P. A. (1988). *Cell*, **52**, 415.
13. Smith, K. C. (1969). *Biochem. Biophys. Res. Commun.*, **34**, 354.
14. Lin, S.-Y. and Riggs, A. D. (1974). *Proc. Natl. Acad. Sci. USA*, **71**, 947.
15. Hillel, Z. and Wu, C.-W. (1978). *Biochemistry*, **17**, 2954.

A2

Suppliers of specialist items

Amersham International plc, White Lion Road, Amersham, Bucks HP7 9LL, UK.

Applied Biosystems, Birchwood Science Park North, Warrington, Cheshire WA3 7PB, UK.

Beckman Instruments Inc., 2500 Harbor Blvd., Fullerton, CA 92634, USA.

Bio-Rad Laboratories, 3300 Regatta Blvd, Richmond, CA 94804, USA.

Boehringer-Mannheim, Sandhoferstrasse 116, Postfach 310120, D6800, Mannheim, Germany.

British Drug House (BDH) Ltd., PO Box 8, Dagenham, Essex RM8 1RY, UK.

Calbiochem, PO Box 12087, San Diego, CA 92112, USA.

Clontech Laboratories, 4030 Fabian Way, Palo Alto, CA 94303, USA.

Fisons, Bishop Meadow Road, Loughborough, Leics LE11 0RG, UK.

Gibco/BRL Life Technologies Inc., 3175 Staley Road, Grand Island, NY 14072, USA.

ICN Flow, 3300 Hyland Avenue, Costa Mesa, CA 92626, USA.

Eastman-Kodak Co., 343 State Street, Bldg 701, Rochester, NY 14652, USA.

New England Biolabs Inc., 32 Tozer Road, Beverly, MA 01915, USA.

Nunc, 2000 North Aurora Road, Naperville, IL 60653, USA.

Perkin Elmer Cetus, 761 Main Avenue, Norwalk, CT 08659, USA.

Pharmacia/LKB, PO Box 175, Björkgarten 30, 75182 Uppsala, Sweden; Piscataway, NJ, USA.

Pierce, 3747 North Meridian Road, PO Box 117, Rockford, IL 61105, USA.

Promega Biotec, 2800 Woods Hollow Road, Madison, WI 53711, USA.

Schleicher and Schull, 10 Optical Avenue, Keene, NH 03431, USA.

Sigma Chemical Co., PO Box 14509, St. Louis, MI 63178, USA.

Stratagene, 11099 North Torrey Pines Road, La Jolla, CA 92037, USA.

Worthington Biochemical Corp., Halls Mill Road, Freehold, NJ 07728, USA.

Index

Index

Index